An Osteology
of Some Maya Mammals

Papers of the Peabody Museum of Archaeology and Ethnology
Harvard University
Volume 73

An Osteology
of Some Maya Mammals

Stanley J. Olsen

Peabody Museum of Archaeology and Ethnology
Harvard University, Cambridge, Massachusetts
1982

Distributed by Harvard University Press

A current list of all Peabody Museum publications available can be obtained

by writing to Harvard University Press

79 Garden Street, Cambridge, Massachusetts 02138

Contents

Illustrations

Acknowledgments

My first acknowledgment of appreciation is to my son, John, a pleasant and always reliable field companion, who is responsible for assembling much of the material that was used in this project. As always, my wife, Eleanor, aided in countless ways with the collecting and processing of comparative material. I wish to thank Charles O. Handley, Jr., of the Division of Mammals of the United States National Museum, Richard Van Gelder of the Department of Mammalogy of the American Museum of Natural History, and Barbara Lawrence of the Mammal Department of the Museum of Comparative Zoology, Harvard University, for the generous loan of skeletal material needed for this study. Without the liberal assistance of Ms. Ann Fallon, this report would not have seen the light of day for some time because of the ever spiraling cost of production. I am truly indebted to her for her kindness. I am grateful to the National Science Foundation for their generous Grant No. BNS-8019400, which helped to defray the costs of publishing this volume. Last but not least is my gratitude to Ms. Donna M. Dickerson for the careful and painstaking task of editing a not altogether clear manuscript.

Stanley J. Olsen

Mammals from the Maya Area

Order Marsupialia
    Family Didelphidae
        Philander opossum, four-eyed opossum
        Marmosa mexicana, Mexican mouse opposum
        Caluromys derbianus, woolly opossum

Order Insectivora
    Family Soricidae
        Cryptotis micrura, Guatemalan small-eared shrew

Order Chiroptera
    Family Emballonuridae
        Rhynchonycteris naso, Brazilian long-nosed bat
    Family Noctilionidae
        Noctilio leporinus, Mexican bulldog bat
    Family Phyllostomidae
        Artibeus lituratus, big fruit-eating bat
    Family Desmodontidae
        Desmodus rotundus, vampire bat
        Diphylla ecaudata, hairy-legged vampire bat
    Family Natalidae
        Natalus mexicanus, Mexican funnel-eared bat
    Family Vespertilionidae
        Eptesicus fuscus, big brown bat
    Family Molossidae
        Molossus rufus, red mastiff bat

Order Primates
    Family Cebidae
        Alouatta villosa, howler monkey
        Ateles geoffroyi, Geoffroy's spider monkey

Order Edentata
    Family Myrmecophagidae
        Myrmecophaga tridactyla, giant anteater
        Tamandua tetradactyla, tamandua
        Cyclopes didactylus, two-toed anteater
    Family Bradypodidae
        Bradypus griseus, three-toed sloth
        Choloepus hoffmanni, two-toed sloth

Order Rodentia
    Family Geomyidae
        Heterogeomys hispidus, hispid pocket gopher
    Family Heteromyidae
        Liomys salvini, Salvin's spiny pocket mouse
    Family Cricetidae
        Oryzomys couesi, Coues' rice rat
    Family Erethizontidae
        Coendou mexicanus, Mexican porcupine
    Family Dasyproctidae
        Agouti paca, paca
        Dasyprocta punctata, agouti

Order Carnivora
    Family Procyonidae
        Potos flavus, kinkajou
    Family Mustelidae
        Eira barbara, tayra
        Galictis allamandi, grison
        Spilogale angustifrons, southern spotted skunk
        Lutra canadensis, northern river otter
        Lutra annectens, southern river otter

Order Sirenia
    Family Trichechidae
        Trichechus manatus, manatee

Order Perissodactyla
    Family Tapiridae
        Tapirus bairdii, Baird's tapir

Order Artiodactyla
    Family Tayassuidae
        Tayassu tajacu, collared peccary
        Tayassu pecari, white-lipped peccary
    Family Cervidae
        Odocoileus (=Dama) virginianus, white-tailed deer
        Mazama americana, red brocket deer

Introduction

Since this publication is designed primarily for use by archaeologists and not by zoological specialists, the animals chosen for discussion and illustration and the method of presenting the data differ from standard osteological texts.

Some mammals present in archaeological sites in the Maya area are not included here because they were included in an earlier discussion of North American forms (Olsen 1964). The southern limits of their ranges extend into the Maya area, so these animals must be considered in any site evaluation of the area. These include the rabbit *(Sylvilagus audubonii)*, wolf *(Canis lupus)*, coyote *(Canis latrans)*, and jaguar *(Felis onca)*. Although the white-tailed deer *(Odocoileus virginianus)* was also included in the earlier study, it is presented here in a different manner in order to facilitate comparison with a similar, though smaller cervid, the brocket deer *(Mazama americana)*.

Several large groups of mammals that are present both as living animals in the Maya area and as remains recovered from excavations are included: the rodents (Rodentia), the bats (Chiroptera), and to a lesser degree, the moles and shrews (Insectivora). All of these animals occur as intrusive forms in Maya sites because of their living habits (burrowing in the case of rodents and moles or living in the fallen structure walls of ruins in the case of bats). Their inclusion is to illustrate the variety of skull forms and dentitions and to aid the field or laboratory worker who may not possess the expertise to recognize the remains of animals of these orders or who may not have adequate comparative skeletons at hand to aid their interpretation.

The selection of illustrations is geared to the majority of archaeologists who do not have trained mammalogists available in the field or in their laboratories to do the preliminary sorting of bones.

As many key characters as possible have been included on the plates and captioned for ready reference. Where possible, dashed lines and arrows have been used in place of notes to indicate diagnostic morphological features. I have found that the value of this style is to cut down on the text, thus facilitating identification in the field and laboratory.

When analyzing excavated bones by comparing them with the elements in the following discussions, the first consideration should be the comparative size of the animal being examined, and size scales have been included on the plates for this purpose.

The goal of this text is to help workers with a limited knowledge of osteology to do a credible job of bone identification. This publication is intended to provide a means to at least preliminarily sort a faunal assemblage so that the final work in the osteology laboratory can be accomplished with the aid of an adequate comparative skeletal collection. In most instances a close determination can be made for those animals having characteristics peculiar to specific mammals by using the keyed illustrations alone. However, sometimes only a taxonomic level of order or family can be deduced without further comparison of actual skeletal material. This is true for most bats and rodents. Thus, an osteological collection of some completeness is a basic requirement for a thorough evaluation of any excavated faunal collection, but it must be supplemented by a text such as this.

1

Skull and Mandible

Order: Marsupialia — Opossums (figs. 1–3)

Several genera of opossums are found in the Maya area. One of these, the Virginia opossum (*Didelphis marsupialis*), has been figured and described in an earlier publication in this series (Olsen 1964). Three others, *Philander*, *Marmosa*, and *Caluromys*, are referred to here. *Didelphis* was reported as occurring in archaeological excavations at Mayapan (Pollock and Ray 1957), at Zaculeu (Woodbury and Trik 1953), at Altar De Sacrificios (Olsen 1972), and at Seibal (Olsen 1978). This latter Maya site also produced remains of *Philander opossum*.

The habits of these small marsupials must always be considered when evaluating their presence at a Maya site. Their remains may be intrusive. Many small animals seek the natural living areas provided by fallen stone structures. Ideal nesting sites are available when interstices are left between large stones that fall haphazardly as a structure wall collapses.

The opossums have a skeleton and, in particular, a skull and dentition that are reminiscent of some early Tertiary mammals. These primitive features are present in the relatively small braincase, high sagittal crest, and sharp-cusped tritubercular molar teeth.

Generally the temporal ridges in opossums unite to form more or less high, thin sagittal crests. The auditory meatus is not fused to the other cranial bones.

The dental formula is generally P_4^5, C_1^1, P_3^3, M_4^4, totaling fifty teeth.

In *Philander* there are two pair of prominent fenestrae in the posterior portion of the palate. The anterior pair are long and ovaloid. The posterior ones are round openings. The nasals are noticeably elongated. There are no postorbital processes. A well-developed sagittal crest is present. P^3 is larger than P^2. M^3 is larger than M^1. The lower margin of the mandible is deflected inward at the angle, terminating in a noticeable point.

In *Marmosa* palatine fenestrae are absent. The nasals are noticeably shortened. There are no prominent postorbital processes. A sagittal crest is lacking. P^2 is larger than P^3. M^1, M^2, and M^3 are the same size. The lower margin of the mandible is deflected inward at the angle, terminating in a noticeable point.

In *Caluromys* there is one anterior pair of irregular, elongated palatine fenestrae. There is no accessory posterior pair of palatine fenestrae, and there are no shortened nasals. The angle of the mandible is slightly turned in. Well-developed postorbital processes are present on the frontals. There is a slight sagittal crest. P^2 is larger than P^3. M^1, M^2, and M^3 are the same size.

The different genera of marsupials found in the Western Hemisphere share many of the same morphological differences that separate them from other contemporary small animals of their area. Size differences, apparent in *Philander*, *Marmosa*, and *Caluromys*, should be noted as one means of distinguishing these similar animals from one another.

Order: Insectivora — Shrews and Moles (fig. 4)
Chiroptera — Bats (figs. 5–12)

The shrews and bats are discussed together not because of any taxonomic similarity but because their occurrence in a Maya site may be due to circumstances that are common to both groups of mammals. Bats and shrews may be present as occupants of destroyed masonry structures rather than because of any interest or use by human inhabitants of the site. Some interest in bats by pre-Columbian cultures in Middle America is evidenced by the bat motif commonly being employed as a vessel decoration.

The small size and delicate structure of both bat and shrew bones also contribute to the rarity of their bones in an archaeological context. Two shrew skulls (*Cryptotis mayensis*) and two bones of the leaf-nosed bat (*Mormoops megalophylla*) were identified from Mayapan (Pollock and Ray 1957). However, they are rarely encountered in an archaeological excavation.

The skeletal structure of bats is greatly specialized because of their adaptation for flight. There is considerable morphological variation among the genera. In the Maya area alone, 75 genera have to be considered when deciding on comparisons (Hall and Kelson 1959). Only eight are referred to here so that the archaeologist may obtain a general idea of their diversified structure. Representative skeletal elements are figured and discussed to enable the field and laboratory worker to broadly identify these small mammals.

The shrew cranium (fig. 4) is conical in form with a tapered, elongated rostral area that is depressed. The nasals and premaxillae are well developed. The occipital area is relatively broad with no sagittal crest. Zygomatic arches are lacking. There are no postorbital processes. Auditory bullae are lacking. The tympanic bones are ring-shaped. The cheek teeth are sharp-cusped with paracones and metacones forming a W-shaped outer wall on the upper molars. The first upper incisors are hooklike and characteristically colored a deep red-brown.

The mandible exhibits an extended angular process. The articular condyle is strongly developed. The first lower incisor is noticeably directed horizontally forward.

Bat skulls show considerable variety in form as evidenced by the small sample selected for inclusion in this study (figs. 5–12).

In general the cranium is comparatively large and rounded. The sense of hearing in bats is highly developed, with this region of the skull being considerably swollen. The tympanic bones are slightly connected with the rest of the skull. The premaxillae are generally small or absent. The orbits are only rarely closed behind. The sagittal crest may be prominent or entirely absent depending on the families or genera being compared. In like manner the shape of the nasals and premaxillae and whether the facial area is compressed or elongated will vary from one family to another. The upper teeth are often triangular or squared with a W-shaped cusp pattern (figs. 5, 6, 10, 11, 12).

In the mandible the coronoid process is rather large. The angle is often rounded (thereby actually negating the morphological term for this area), and the condyle is quite expanded transversally.

The characteristics that define the different illustrated skulls are best referred to in taxonomic order.

In the Brazilian long-nosed bat, *Rhynchonycteris naso* (fig. 5), there is no rostral angle. The premaxillae are short. The postorbital processes are well developed and curved. There is no noticeable sagittal crest. The mandible has a decided upward curve behind the tooth row. A W-shaped cusp pattern is present in the upper molar series.

In the Mexican bulldog bat, *Noctilio leporinus* (fig. 6), the skull has no distinct postorbital processes. The premaxillae have both palatal and nasal branches. The latter are markedly elongate. The palate is complete and closed anteriorly. The braincase is oval in outline, having a distinct sagittal crest dividing anteriorly and extending downward over the tooth row. The mastoid area of the skull flares outward. A W-shaped cusp pattern is present in the upper molars. The comparatively heavy horizontal ramus has a low coronoid process.

The big fruit-eating bat, *Artibeus lituratus* (fig. 7), has a moderately inflated, high-domed skull with a well-defined sagittal crest. The zygomatic arches are widespread and short. The rostrum is low and set off from the cranium by a slight angle. The wide palate ends in a restricted, square posterior extension. The upper molars are of a crushing type, having a finely crenulated surface. The horizontal ramus is heavily

constructed with a low ascending ramus and coronoid process.

The vampire bat, *Desmodus rotundus* (fig. 8), can best be identified by its specialized dentition, used for making an incision in the skin of its victim in order to obtain the blood supply which comprises its diet. The sharp sickle- or lancet-shaped teeth are composed of the first upper incisors and the lower canines. All traces of crushing teeth are absent in the vampire. The braincase is quite expanded posteriorly, forming an ovaloid cranial vault with no defined sagittal crest. There is almost no defined rostral area. The auditory bullae are quite expanded. The posterior region of the mandible is quite wide with little or no visible coronoid process. There is a stepped angle on the ventral margin of the horizontal ramus below the posterior end of the tooth row.

The hairy-legged vampire bat, *Diphylla ecaudata* (fig. 9), has proportionately smaller upper incisors. The bullae are considerably larger than in *Desmodus*. The braincase is also somewhat more expanded and extends anteriorly over the upper canines where it takes a decided downward angle. There is no sagittal crest. The posterior segment of the mandible is weaker than that found in *Desmodus*.

The Mexican funnel-eared bat, *Natalus mexicanus* (fig. 10), has a high-domed braincase that has a decided rostral angle and a much elongated muzzle. The upper molar cusp pattern is W-shaped. The mandible has a noticeably long, thin horizontal ramus and a low ascending branch.

The big brown bat, *Eptesicus fuscus* (fig. 11), and other members of this genus are, perhaps, among the most common bats observed in the Maya area. The skull has a noticeably slanting braincase with its apex at the anterior margin. There is a considerable break or angle at the rostrum, which is flattened. There is a low sagittal crest. There are no postorbital processes on the skull. The palate is wide and solid and narrows abruptly behind the tooth row. The sides of this extension are nearly parallel. The molars exhibit a W-shaped cusp pattern. The upper incisors are well developed, as are the canines. There is a slight diastema between the lateral incisors and the canines. There is a hooked process extending from the mandible at the angle of the ascending ramus.

The red mastiff bat, *Molossus rufus* (fig. 12), has a skull that is squared at the occipital end. It is restricted in the rostral area, forming an hourglass shape when viewed dorsally. The sagittal crest is well developed. There are no postorbital processes. The palate ends at the posterior end of the tooth row. Upper canines are well developed; upper molars exhibit a W-pattern on the occlusal surfaces. The lower jaw is strongly constructed, having a hooked process extending at the angle. There is a low ascending ramus.

Order: Primates — Monkeys *(figs. 13, 14)*

In general the skull vaults of monkeys are inflated and rounded. The orbits are directed anteriorly and separated from the temporal fossae by a bony bar or partition. The incisors are chisel-shaped. The lower canines are pointed. There is little inflation of the auditory bullae.

Six small primates are indigenous to Central America. Only two, the howler monkey (*Alouatta villosa*) and the spider monkey (*Ateles goeffroyi*), are included in this study.

The cranium of *Alouatta villosa* (fig. 13) is slightly compressed dorsoventrally with a squared occipital margin when viewed dorsally. There is a noticeable projecting inion. The zygomatic arches are heavy and flare away from the sides of the skull. The muzzle is square and shortened. The canines, premolars, and molars are strongly constructed. The mandible is noticeably deep in the region of the angle.

The cranium of *Ateles geoffroyi* (fig. 14) has a vault that is high and rounded. The zygomatic arches are weakly developed and lie close to the sides of the skull. The upper canines, premolars, and molars are of moderate size. The incisors project forward and away from the premaxillae. The mandible is lightly built.

Order: Edentata — Anteaters and Sloths *(figs. 15–19)*

The anteaters, as the name of the order implies, are edentulous. As a group, their skulls are readily identified if complete, with virtually no comparison with other forms.

The skull of the giant anteater, *Myrmecophaga tridactyla* (fig. 15), is distinctive in its large size as well as in the unusual structure that is associated with the anteater cranium. The skull is smooth, evenly rounded, long, and tapering. The rostrum is tubular, round above, and flattened below. It is composed chiefly of the maxillae and the nasals. The premaxillae are small. The prelacrimal region of the skull is more than twice the length of the postlacrimal region. The pterygoids are joined, prolonging the narial passage for almost the entire length of the skull. The zygomatic arch is incomplete. The jugals are attached to the maxillae. The bullae are compressed. There is a large occipital condyle. The mandible is long and slender with a slight, backward-curving coronoid process.

The skull of the tamandua, *Tamandua tetradactyla* (fig. 16), is similar in general structure to that of the giant anteater except for its smaller size and the cranial/rostral proportions. The prelacrimal region of the skull is shorter than the postlacrimal region. The pterygoids are united and prolong the narial pasage for almost the entire length of the skull. The zygomatic arch is lacking. The jugals are attached to the maxillae. The bullae are compressed. The mandible is slender with a slight, backward-extending coronoid process.

The two-toed anteater, *Cyclopes didactylus* (fig. 17), is the smallest of the three anteaters presented in this study. When viewed laterally, the cranium has a decided convex profile and is greatly shortened. The pterygoids and the posterior portions of the palatines do not meet below the posterior nares to form a tube as in the others. There is an open channel instead. The zygomatic arch and the jugal are absent. The premaxillae are vestigial. The bullae are moderately inflated. The mandible has a prominent, narrow, recurved coronoid process and a well-developed angular process that is equal in length to the coronoid process. The longitudinal ramus is curved downward at its anterior end.

In general the sloths have skulls in which the separate sutures are obliterated or fused early in the animals' development. The facial areas are short and blunt.

The three-toed sloth, *Bradypus griseus* (fig. 18), has teeth that are subcylindrical and peglike. The anterior tooth in the maxilla is smaller than the succeeding one. The highest point on the skull is located anteriorly and directly above the orbits. The anterior nares are nearly vertical. The premaxillae are rudimentary. The palate is narrow and does not extend backward beyond the limit of the tooth row. The postorbital process is small. The zygomatic arch is incomplete. The jugal terminates in two widely separated branches. The lower jaw terminates bluntly at the symphysis. The angle extends beyond the limits of the articular condyle in a wide, flattened process.

The two-toed sloth, *Choloepus hoffmanni* (fig. 19), differs in a number of areas from its three-toed relative. The highest point on the skull is at the midpoint on the cranium. The teeth are simple cylindrical or peglike structures. The anterior teeth above and below are the largest. They are caniniform, having sharp, beveled cutting edges. The premaxillae are more developed than in *Bradypus*. The zygomatic arch is incomplete. The jugal terminates in a flaring bifurcate branch. The lower jaw has a decided forward-projecting spoutlike symphysis. The angle and articular condyle extend about the same distance to the rear of the mandible.

Order: Rodentia — Rodents *(figs. 20–25)*

The rodents of the Maya area are varied in both size and form. As with the insectivores and chiropters, some may occur intrusively in archaeological sites because of their natural burrowing habits. As with the bats, only a small sample of those occurring in the area have been selected to familiarize the archaeologist with the general osteological structure of the order as well as to aid in the positive identification of some of the more distinctive species.

In general rodents are gnawing animals. The huge incisors, two above and two below, are quite characteristic of these mammals. These teeth are generally heavy, chisel-shaped, and admirably suited for cutting. They are ever growing and kept continually sharp by close occlusion between the upper and lower pair. The lateral incisors, the canines, and the anterior premolars are lacking. There is a wide gap or diastema between the anterior incisors and the cheek teeth. The types of molars and the patterns on their grinding surfaces, formed by the arrangement of the soft dentine and hard enamel, are consistent among the different rodent groups and are therefore of value for classification and identification even when the skeletal evidence is quite fragmentary.

Most rodents have tooth structures that are one of two general types. They can be either brachydont (low-crowned) or hypsodont (high-crowned). In the brachydont dentition, only the roots occur in the bony sockets or alveoli, leaving the crowns of the teeth fully exposed. In the hypsodont teeth, not only the roots but part or much of the crown may be embedded in a single alveolus below the bony margin of the jaw. This latter type has also been termed a pillar-type molar. In both types, a general classification can be arrived at by observing only the alveoli, even when the teeth are entirely absent.

General discussions about this group of animals as they occur in Central America are not found in popular volumes on mammals. Two publications for those who wish to analyze this group in more detail have been published by Leopold (1959) and Walker (1975).

The pocket gophers, of which *Heterogeomys hispidus* (fig. 20) is a typical example, are burrowers which dig quite deeply into the soil. This species occurs in Maya excavations. Pollock and Ray (1957) identified remains of this species from excavations at Mayapan.

The upper incisors are diagnostic in pocket gophers. In general they have grooves on the anterior surfaces which run along the entire outer surface of the enamel face. The number and location of these grooves or sulci are of taxonomic value in identifying the rodent in question. They may be single or double and centrally located or on the medial or lateral side of the midline of the tooth. In *Heterogeomys hispidus*, the grooves are located on the medial side of the midline. As with other rodents, there are no canines. There are wide diastemas btween the incisors and cheek teeth.

The skull is wide and flattened on the dorsal surface, forming a squared contact with the occipital area. The sagittal crest is wide but blunt and lies close to the cranial vault. The rostral area is noticeably constricted in the region of the posterior limits of the nasals and premaxillae. The mastoid area projects laterally away from the sides of the skull. The molars are comprised of pillared prismatic plates having subcircular grinding surfaces.

The lower jaw has a characteristic rocker-shaped lower margin. The angles extend laterally away from the ramus. The posterior ends of the horizontal ramus and ascending ramus are wide and heavy. There is a defined coronoid process.

Salvin's spiny pocket mouse, *Liomys salvini* (fig. 21), has cheek teeth that are rooted but high-crowned. The skull is long and slender with forward-projecting nasals and premaxillae. The zygomatic arches are weakly constructed. The bullae are slight and do not extend below the level of the grinding surfaces of the cheek teeth. The mandible has a noticeable extension to the angle. The coronoid process is barely developed and is below the heavily constructed articular condyle. The lower incisors are well developed. The upper incisors are stout but not greatly extended beyond the margins of the premaxillae.

Coues' rice rat, *Oryzomys couesi* (fig. 22), was identified at Mayapan (Pollock and Ray 1957) from midden refuse that found its way into a burial chamber. Pollock and Ray believed that the presence of these rats was due to their scavenging in the midden. The rice rat skull is rather large with a substantial braincase. The rostrum is short, having trenchant lateral margins that project as supraorbital ridges. The nasals and premaxillae do not noticeably project anteriorly beyond the limits of the upper incisors. The bullae are not noticeably inflated. The zygomatic arches are thinly constructed. The palatine vacuities are long oblong slits. The coronoid process is thin and not greatly extended above the articular condyle. The angle of the mandible is noticeably extended (broken away in the illustrated example, fig. 22).

The Mexican porcupine, *Coendou mexicanus* (fig. 23), has a heavy, robust skull with a much broadened frontal region. It is strongly arched in lateral profile, with the highest point above the orbits. The external nares abruptly end at a right angle behind the projecting premaxillae and upper incisors. The bullae are long and narrow and extend downward beyond the limits of the occlusal surfaces of the upper molars. The zygomatic arches are dorsoventrally wide. The interorbital foramen is larger than the foramen magnum. The enamel of upper and lower molars has an infolded pattern that is characteristic of porcupines. The mandible has a decided step-notching of the lower margin. The angle is extended. The coronoid process is low. The heavy articular condyle extends above the coronoid process. Lower incisors are long and heavily constructed; upper incisors are heavily constructed but are not greatly extended beyond the margins of their alveoli. This porcupine has been identified from the Maya site of Seibal (Olsen 1978).

The largest rodents in the Maya area, the paca (Agouti paca) and the agouti (Dasyprocta punctata), have been recorded at several Maya sites. Pollock and Ray (1957) recorded the paca at Mayapan, although they incorrectly listed it as Cuniculus paca (p. 641). They also recorded remains of the agouti. Both rodents were also identified at Seibal (Olsen 1978). I believe their large size and possible food value may be responsible for their presence in Maya excavations. However, recorded evidence associated with the above finds does not indicate this.

The paca, Agouti paca(fig. 24), has a uniquely developed cheekplate composed of the jugal and part of the maxilla. The expanded surface is heavily sculptured, particularly in the adult animal. The dorsal surface of the skull is flattened, with a decided downward curve beginning in the rostral area and continuing to the anterior end of the nasals. The temporal and sagittal crests are also sculptured ridges. The mastoid area projects slightly beyond the limits of the lateral margin of the skull. The molar teeth have prominent reentrant folds that wear into isolated, narrow, enamel-rimmed lakes. The lateral surface of the mandible, particularly in the area of the angle, has a deeply sculptured surface that is diagnostic for this species. The mandible has a narrow extension to the anterior end, from which a long, thin incisor extends. The lower margin of the longitudinal ramus has a decided step below the last molar. There is no extension to the angle, which is heavy and deep. There is no noticeable coronoid process. The articular condyle is situated on a low, rounded termination of the ascending ramus.

The agouti, Dasyprocta punctata (fig. 25), has a rather slender skull with a curving outline when viewed laterally. There is little or no sagittal crest. The auditory bullae are moderately large. The palatine foramina are short and anteriorly situated. The zygomatic arches are rather weakly constructed. The paroccipital processes are prominent. The anterior end of the skull is narrow and noticeably extended. The angle of the mandible is extended backward, causing a decided break in the lower margin of the jaw. Neither the coronoid process nor the articular condyle is noticeably produced. The hypsodont molariform teeth have a definite pattern formed by the infolded, enamel grinding surfaces.

Order: Carnivora — Procyonids and Mustelids
(figs. 26–31)

The skull of the kinkajou, Potos flavus (fig. 26), is short and rounded. The cranium is highly arched and constricted in the postorbital region. The rostrum is short and broad. The postorbital processes are well developed. There is no sagittal crest. The palate is heavy with small fenestrae at the anterior end. The auditory bullae are flattened. The molars are low-crowned. Both upper and lower canines are well developed. The mandible is heavily constructed with a high ascending ramus. The angle has a decided downward curve, which begins about midway on the horizontal ramus. The symphysis is strong and heavily constructed. It is doubtful that this procyonid will be commonly encountered in an archaeological context within its range, but it is included as an example of this group. The raccoon, cacomistle, and coati (members of the same family) were described and figured in an earlier publication in this series (Olsen 1964).

The tayra, Eira barbara (figs. 27, 28), has a skull that is characteristic of mustelids in general. The cranium is rounded but not expanded. There are double sagittal crests that are diagnostic for this species. It is comparatively long, with the orbits and rostrum occupying the forward one-third of the skull. The zygomatic arch is comparatively thin and weak. The palate is solid, with two small, oval fenestrae located between the canines. There are two widely separated sagittal crests running parallel from the occipital region and flaring out as postorbital processes. The bullae, as in most mustelids, are flattened and invisible when the skull is viewed from a lateral aspect. The upper and lower canines are well developed. The premolars and the carnassials are equipped with sharp cusps and cutting surfaces. The last upper molar is constricted at its middle, creating an hourglass-shaped tooth.

One of the most interesting artifacts recovered from the Maya site of Seibal was an amulet or possibly a lime container made from the pierced cranium of a tayra (fig. 28 and Olsen 1978, fig. 173). The two holes used for suspending the object were considerably worn along their edges, indicating some extended length of use. The cranium had been considerably broken when recovered, and there is no other evidence to suggest a more plausible use for this pierced skull fragment.

The other large mustelid, one of a similar size to the tayra, is the grison, Galictis allamandi (fig. 29). The grison skull is quite similar to that of the tayra in most aspects. The most striking difference is that the grison has a single, low sagittal crest. The cranium also has two occipital crests that project from the back of the skull, forming shelflike angles that end just short of the zygomatic arch. This same area in the tayra is rounded in outline. The zygomatic arch is considerably more massive in the grison. The mandibles in both the tayra and grison are heavily constructed. The coronoid apex terminates in a rounded margin in the tayra. In the grison it is a blunted angle projecting downward and backward.

The southern spotted shunk, *Spilogale angustifrons* (fig. 30), has been reported from a cenote deposit in Mayapan (Pollock and Ray 1957). The typical mustelid skull is mostly composed of braincase. The orbits and rostrum comprise the anterior one-third. The sagittal crest is nearly lacking. The zygomatic arches are thin and weakly constructed. Two round fenestrae pierce the palate between the canines. The occipital margin is square or slightly dished when viewed dorsally. The bullae are considerably flattened. The canines are well developed. The premolars and molars are sharply cusped with prominent carnassial teeth. The last upper molar is constricted at midpoint. The mandible is strongly built and has a rather high ascending ramus with a rounded apex.

The southern river otter, *Lutra annectens,* is considerably larger than the northern river otter, *Lutra canadensis* (fig. 31). Their ranges do not overlap (Hall and Kelson 1959, p. 947, map 477), but I have found that comparative material of the southern species is scarce in all but the larger museum osteology collections. The northern form is quite common. Both skulls have been illustrated so that it may be possible for an archaeologist to establish the presence of the genus *Lutra* in the Maya area using the northern species for comparison with the other faunal material.

Remains of *Lutra annectens* were recovered from the Maya site of Altar de Sacrificios (Olsen 1972). The skull of *Lutra annectens* is convex when viewed laterally. It is noticeably constricted between the zygomatic arches. There is no noticeable sagittal crest. The bullae are flattened. The zygomatic arches are moderately strong. The skull has a short, square rostral area. The canines are well developed as are the premolars and molars. The last upper molar is not as constricted in the middle as are those of the other mustelids discussed in this report. The carnassials are prominent and well developed. The mandible is strongly constructed with a pronounced curve to the lower margin. The ascending ramus is moderately developed. It terminates in a rounded apex in the southern form and is more angular in the smaller northern species.

Order: Sirenia — Manatees *(figs. 32, 33)*

The manatee, *Trichechus manatus* (figs. 32, 33), is unique in size and general structure, so there is little problem identifying this animal even when found in a quite fragmentary condition.

A carved manatee rib was recovered at Mayapan (Pollock and Ray 1957).

The bones of the skull and postcranial skeleton of the manatee are dense and massive in structure. There is very little cancellous structure to the bone. It has instead an ivorylike consistency termed pachyostosis. The nasals are vestigial. The rostrum is narrow and pointed. The anterior portions of both the skull and mandible lack teeth. In life, these areas are covered by hornlike plates. The cranial roof is comparatively narrow in relation to the total skull size. The bony orbits are tubelike. The zygomatic arches are heavily constructed. The palate is long and narrow with a single opening at its anterior end. The bullae are weakly attached to the skull. The rooted cheek teeth have parallel cusps not unlike those found in the tapir, *Tapirus bairdii.* The mandible is turned down at the symphysis at an extreme angle. The coronoid process is deflected forward at a considerable angle as is the articular condyle. Generally sutures are weakly united regardless of the age of the individual. Size alone will aid in identifying this animal. Only the tapir approaches it in size in the Maya area.

Order: Perissodactyla — Horses and Tapirs *(figs. 34, 35)*

Baird's tapir, *Tapirus bairdii* (figs. 34, 35), is the only perissodactyl found that is native to the Maya area. The other perissodactyl in North America, the horse, had disappeared from the scene at the close of the Pleistocene period, not to reappear until introduced by the Spanish early in the sixteenth century.

The tapir skull has a high, thin sagittal crest with a decided convex curve when viewed laterally. The cranium has a decided downward rostral curve continuing on to a rather long and thin muzzle, also terminating in a downward curve. The nasals are short and triangular. There is a wide diastema between the upper and lower canines and the cheek teeth. There are prominent grooves alongside the nasals for attachment of the distinctive proboscis of this animal. The muzzle is narrow and constricted when viewed dorsally. Incisors are strong and cupped. Molars and premolars are rooted with diagnostic parallel cusps running transversally and are similar in general form to those found in the manatee. The canines and third incisors are well developed. The mandible is heavily built with a well-rounded angle, a moderately developed coronoid process, and a strong articular condyle. This animal is approached in size only by the manatee.

Order: Artiodactyla — Peccaries and Deer *(figs. 36–43)*

Both the collared peccary, *Tayassu tajacu* (figs. 36, 37), and the white-lipped peccary, *Tayassu pecari* (figs. 38, 39), presently occur in the Maya area. Remains of both peccaries are abundant in most Maya sites. They were common at Seibal, Altar de Sacrificios, and

Mayapan. The condition of the bone fragments indicates that they were an important food source for the Maya. I have classified both species within the same genus, *Tayassu*, following Hall and Kelson (1959). Michael Woodburne (1968, p. 1) reclassified the collared peccary, *Tayassu tajacu*, to the Genus *Dicotyles*. He stated as his reason, "The morphological distinctions between the two living peccaries have historical significance and it is concluded that two separate genera are represented."

I believe that this is an evolutionary problem and open at least to debate before arriving at the above conclusions. I prefer to assign both peccaries to the same genus, *Tayassu*, as it is presently the generally accepted taxonomic genus for both animals.

Differences that separate the skulls of the collared and white-lipped peccaries are minute, and positive separation of the two is not always possible, particularly when based on the examination of fragmentary material. Perhaps the clearest method of presenting these slight differences, which are only apparent when the skulls are compared to each other, would be with the use of the table below.

The white-tailed deer, *Odocoileus virginianus* (figs. 40, 41, 43), has been figured and compared in an earlier publication (Olsen 1964). It is presented here in order to allow for a separation of elements of the smaller cervid, the red brocket deer or mazama, *Mazama americana* (figs. 42, 43).

Unlike the peccaries, these artiodactyls have no dentition in the premaxillae, having instead a dental pad in life.

Except for size, the cheek teeth of both deer are similar in construction. They are rooted, with a cingulum that is quite visible and an occlusal pattern that suggests the capital letter "W." The lower cheek dentitions are similar in both as are the lower incisors, which are small and spatulate. There are long diastemas between the lower incisors and the cheek teeth, those of the *Mazama* being proportionately smaller. The coronoid process on the mandible of the *Mazama* is noticeably wider than that in the white-tailed deer.

The comparative size of these two deer is generally adequate for their specific separation. The brocket deer is generally at least twenty-five percent smaller than the white-tailed deer in all respects.

Proportionately, the white-tailed deer has a longer and more slender muzzle when viewed dorsally. The orbits are circular, and there are prominent lacrimal pits. The brocket has a rostral area that is blunter and with a shorter muzzle. The zygomatic arches are more widespread in the brocket. There is generally a slight concave or dished margin in the rostral area of the brocket when viewed laterally. The same margin in the white-tailed deer is either straight or slightly convex. Antlers are present on the skull of male animals of both species. They are simple spikelike projections in the mazama and are many-tined in the white-tail.

Skeletal remains of deer are common occurrences in most archaeological sites within the Maya area.

Differences between the Two Peccaries

TAYASSU TAJACU	*TAYASSU PECARI*
(a) Cranium, low, wide, and short	Cranium, high, narrow, and elongate
(b) Dorsal surface of rostrum, narrow and convex	Dorsal surface of rostrum, broad and flat
(c) Supraorbital-nasal canals diverge markedly, then converge anteriorly; canals, well defined anteriorly	Supraorbital-nasal canals do not converge markedly anteriorly; canals, not well defined anteriorly
(d) Canine buttress, large	Canine buttress, small
(e) Narial notch, square	Narial notch, pointed
(f) Width of nasals, constricted between premaxillae	Width of nasals, not constricted between premaxillae
(g) Diastema between canine and first premolar, slight	Diastema between canine and first premolar, wide
(h) Palatal area anterior to first premolar, constricted	Palatal area anterior to first premolar, wide
(i) Teeth, comparatively larger	Teeth, comparatively smaller
(j) Condyle of mandible, in vertical line with rear margin of ascending ramus	Condyle of mandible, directed slightly forward
(k) Overall size, smaller	Overall size, larger

Skull and Mandible

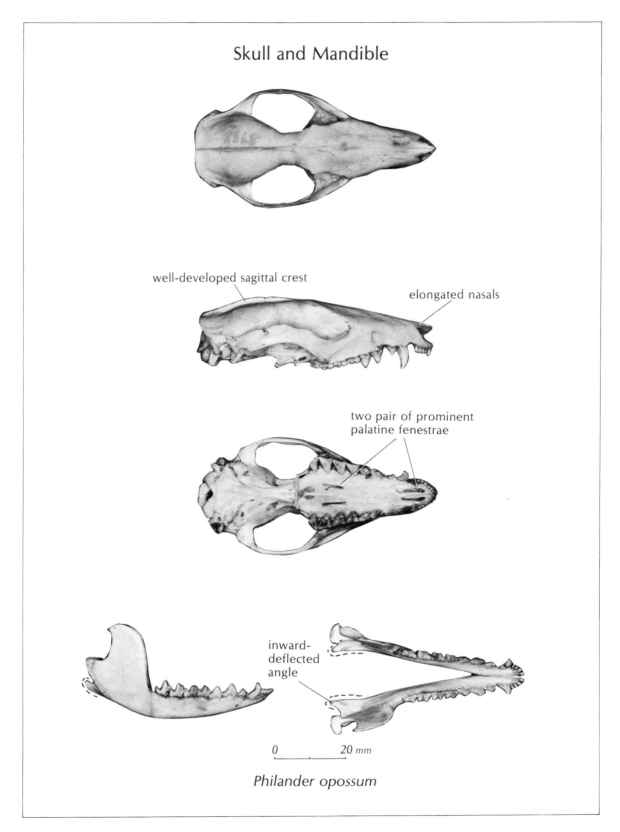

well-developed sagittal crest

elongated nasals

two pair of prominent palatine fenestrae

inward-deflected angle

0   20 mm

Philander opossum

FIGURE 1

Skull and Mandible

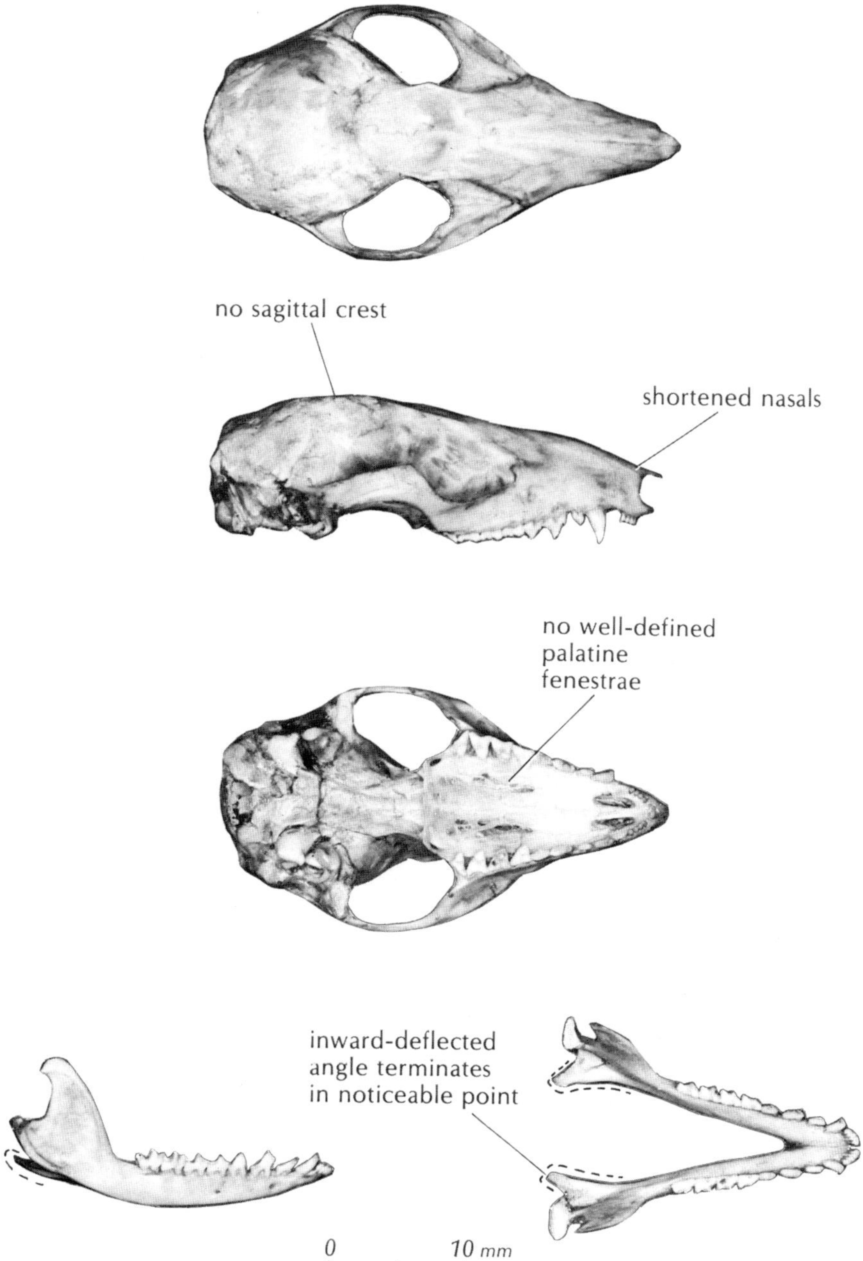

no sagittal crest

shortened nasals

no well-defined
palatine
fenestrae

inward-deflected
angle terminates
in noticeable point

0   10 mm

Marmosa mexicana

FIGURE 2

Skull and Mandible

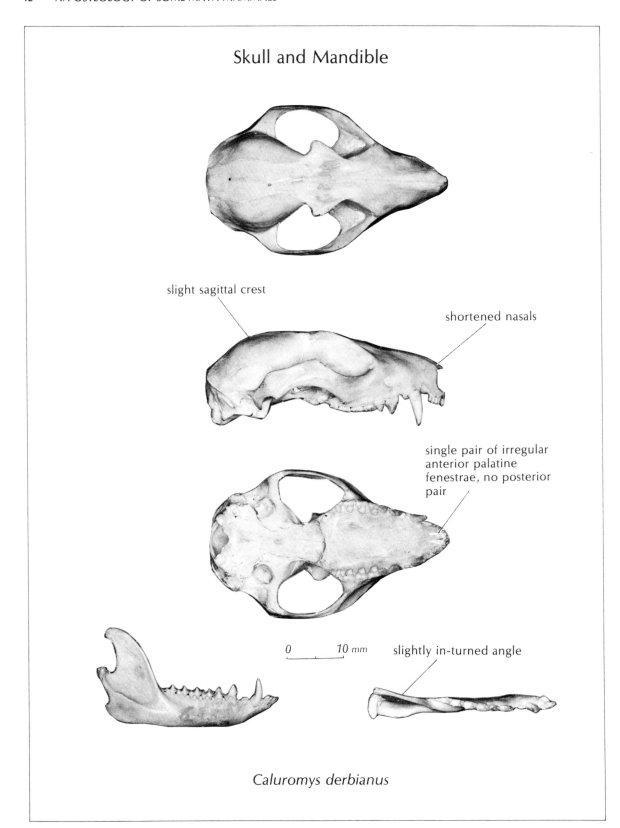

slight sagittal crest

shortened nasals

single pair of irregular anterior palatine fenestrae, no posterior pair

0   10 mm

slightly in-turned angle

Caluromys derbianus

FIGURE 3

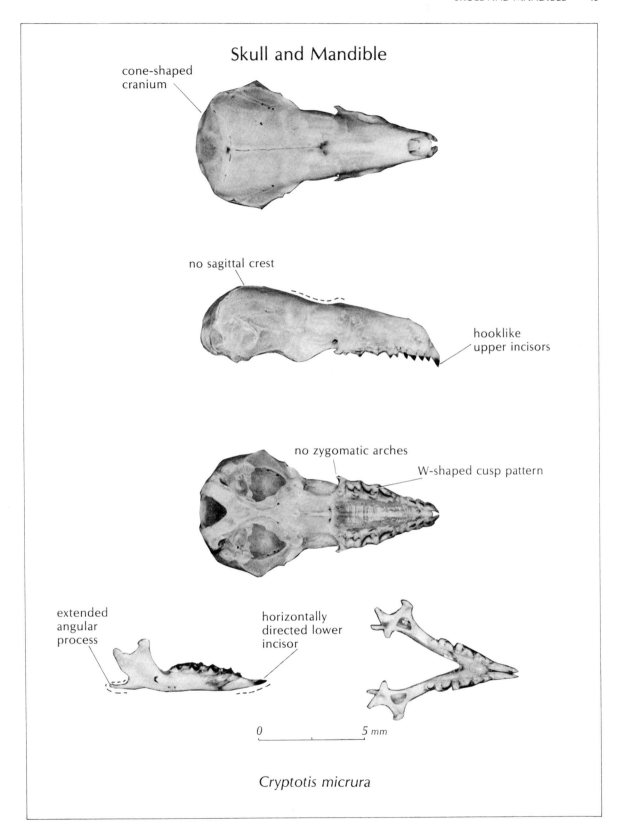

Skull and Mandible

cone-shaped cranium

no sagittal crest

hooklike upper incisors

no zygomatic arches

W-shaped cusp pattern

extended angular process

horizontally directed lower incisor

0   5 mm

Cryptotis micrura

FIGURE 4

Skull and Mandible

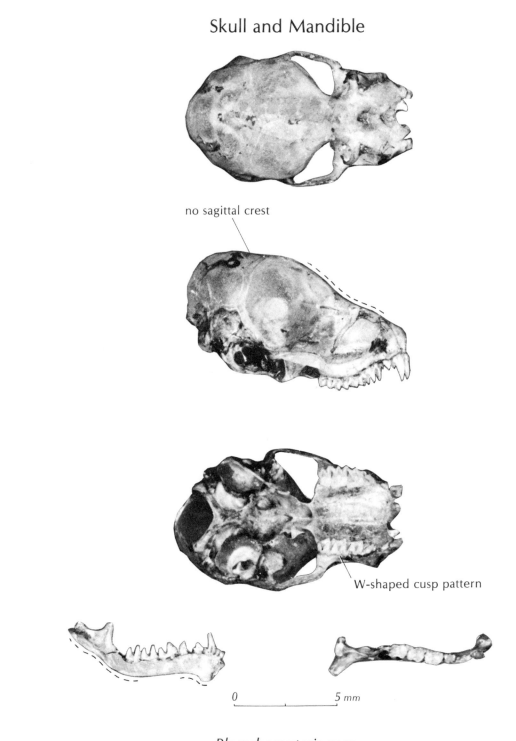

no sagittal crest

W-shaped cusp pattern

0   5 mm

Rhynchonycteris naso

FIGURE 5

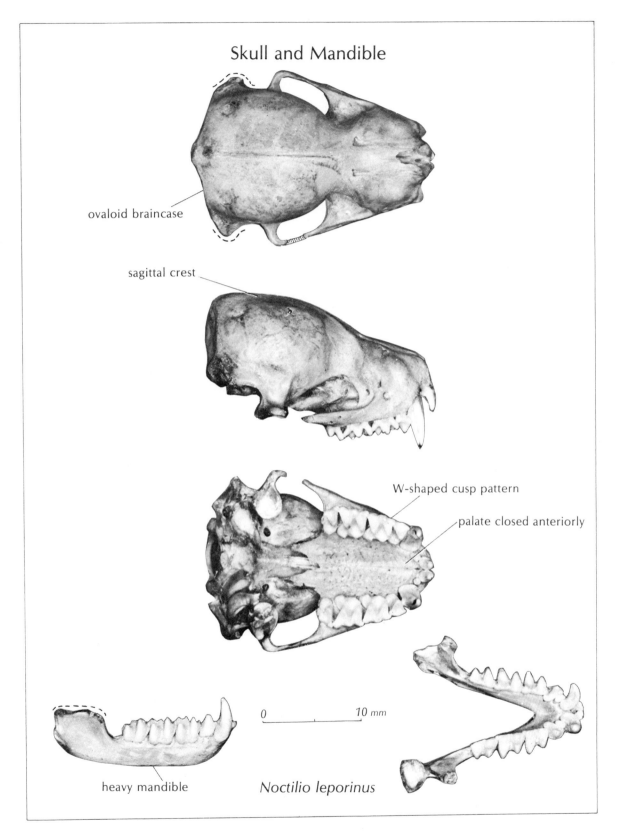

Skull and Mandible

ovaloid braincase

sagittal crest

W-shaped cusp pattern

palate closed anteriorly

heavy mandible

Noctilio leporinus

0   10 mm

FIGURE 6

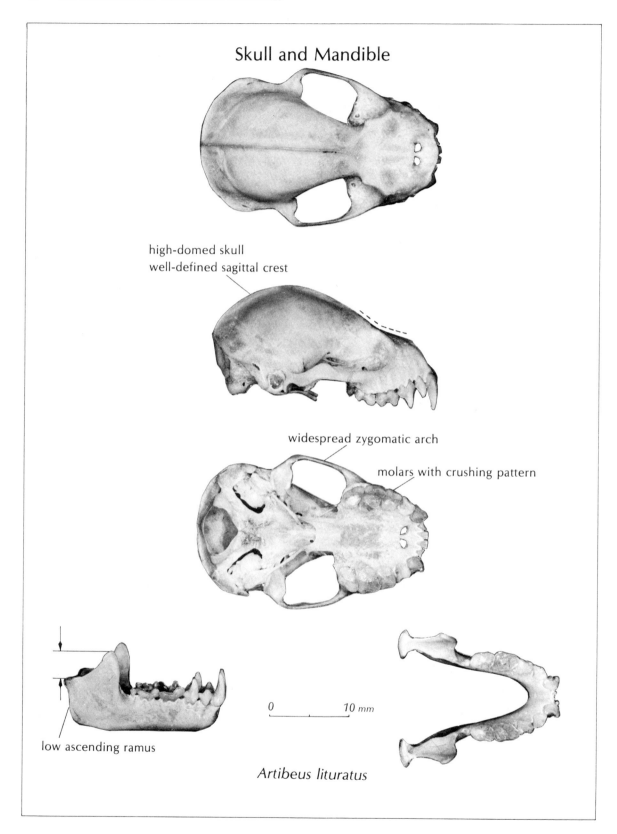

Skull and Mandible

high-domed skull
well-defined sagittal crest

widespread zygomatic arch

molars with crushing pattern

low ascending ramus

0 _____ 10 mm

Artibeus lituratus

FIGURE 7

Skull and Mandible

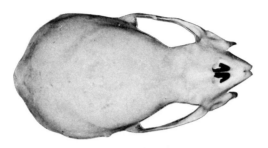

high-domed skull, no sagittal crest

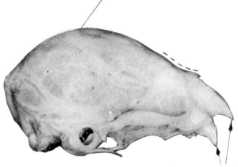

piercing lancet-
shaped teeth

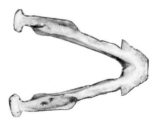

0   10 mm

Desmodus rotundus

FIGURE 8

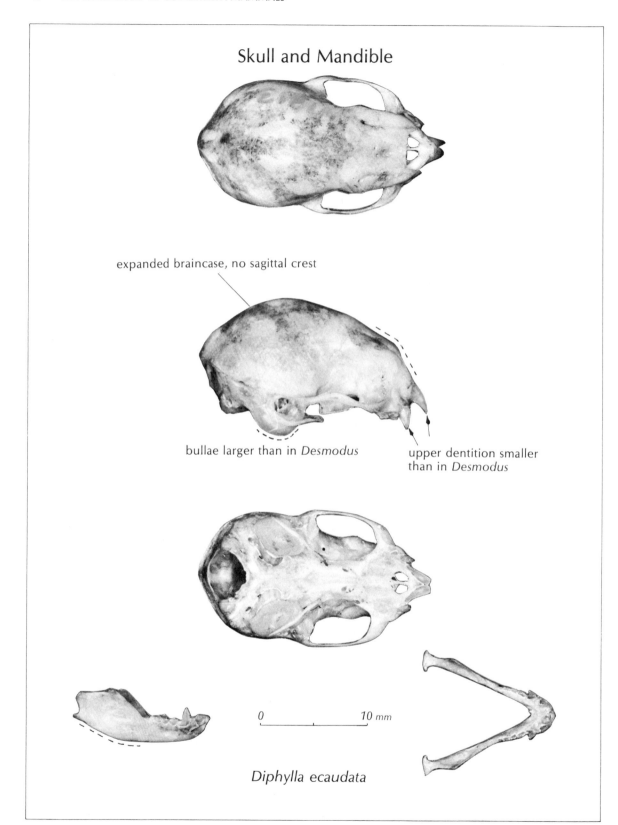

Skull and Mandible

expanded braincase, no sagittal crest

bullae larger than in *Desmodus*

upper dentition smaller than in *Desmodus*

0   10 mm

Diphylla ecaudata

FIGURE 9

Skull and Mandible

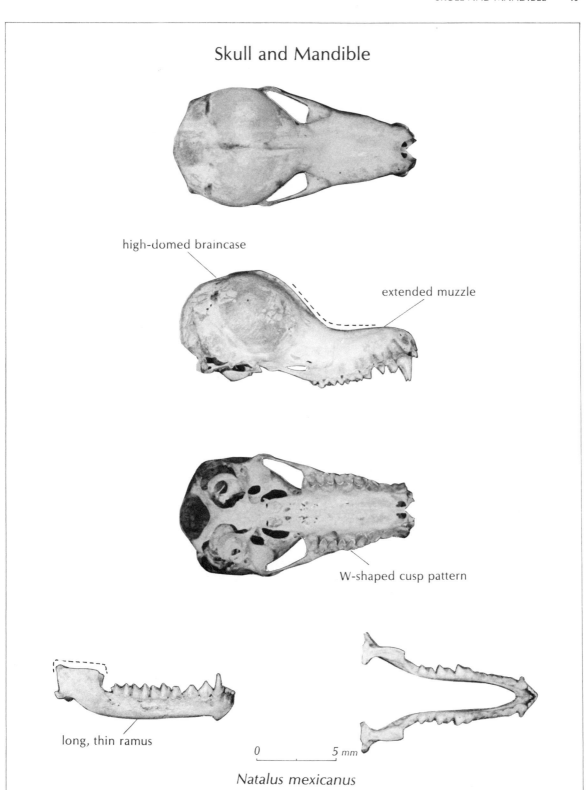

high-domed braincase

extended muzzle

W-shaped cusp pattern

long, thin ramus

0   5 mm

Natalus mexicanus

FIGURE 10

Skull and Mandible

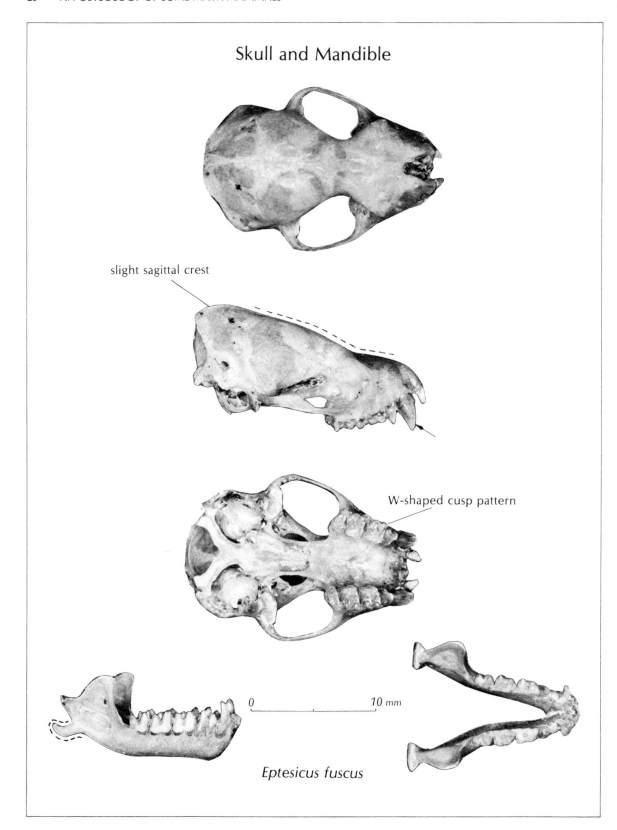

slight sagittal crest

W-shaped cusp pattern

0   10 mm

Eptesicus fuscus

FIGURE 11

Skull and Mandible

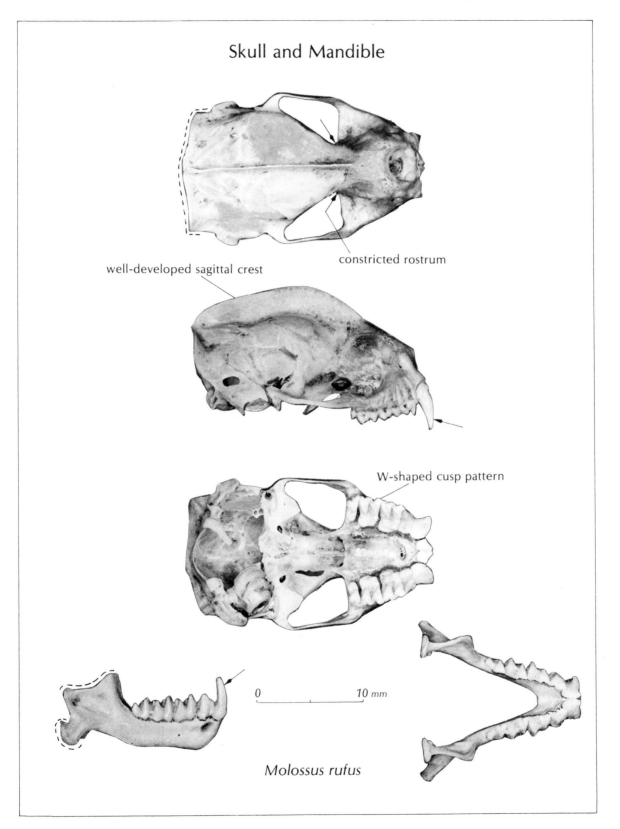

well-developed sagittal crest

constricted rostrum

W-shaped cusp pattern

0   10 mm

Molossus rufus

FIGURE 12

Skull and Mandible

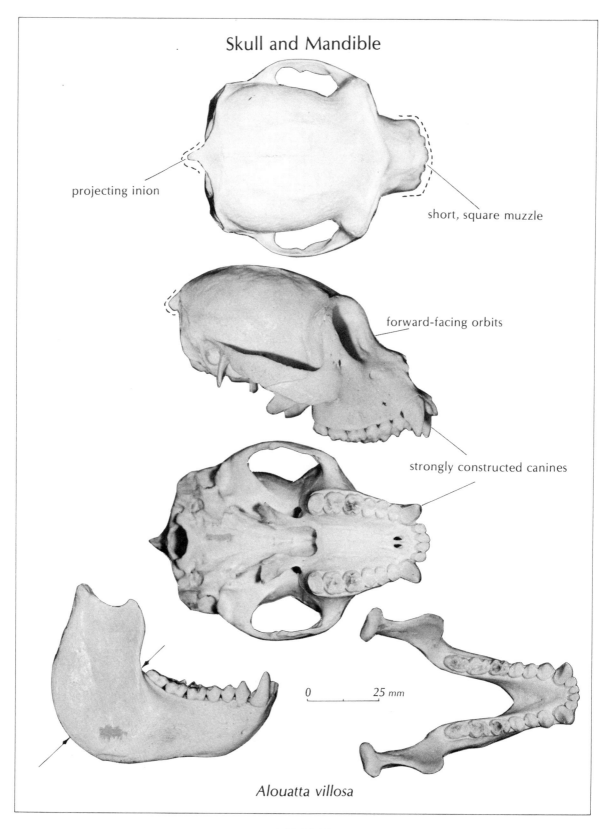

projecting inion

short, square muzzle

forward-facing orbits

strongly constructed canines

0   25 mm

Alouatta villosa

FIGURE 13

Skull and Mandible

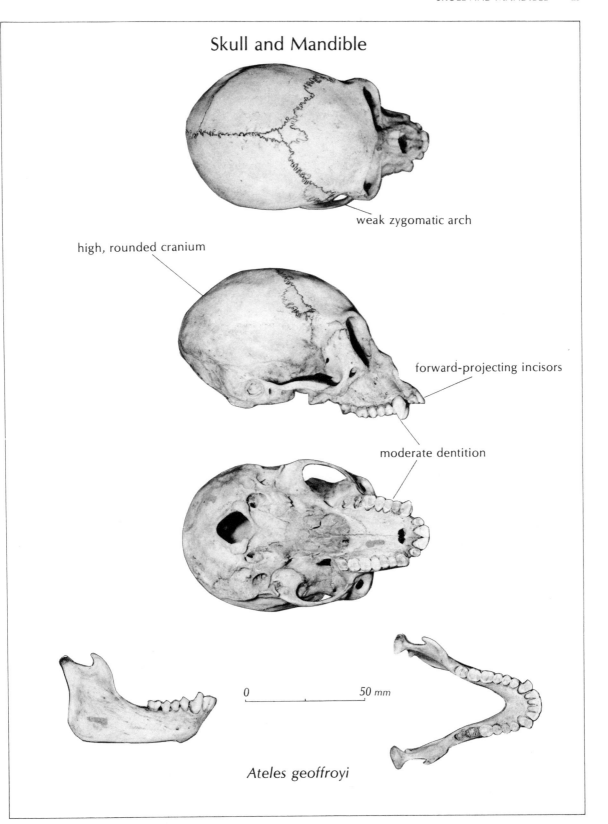

weak zygomatic arch

high, rounded cranium

forward-projecting incisors

moderate dentition

0   50 mm

Ateles geoffroyi

FIGURE 14

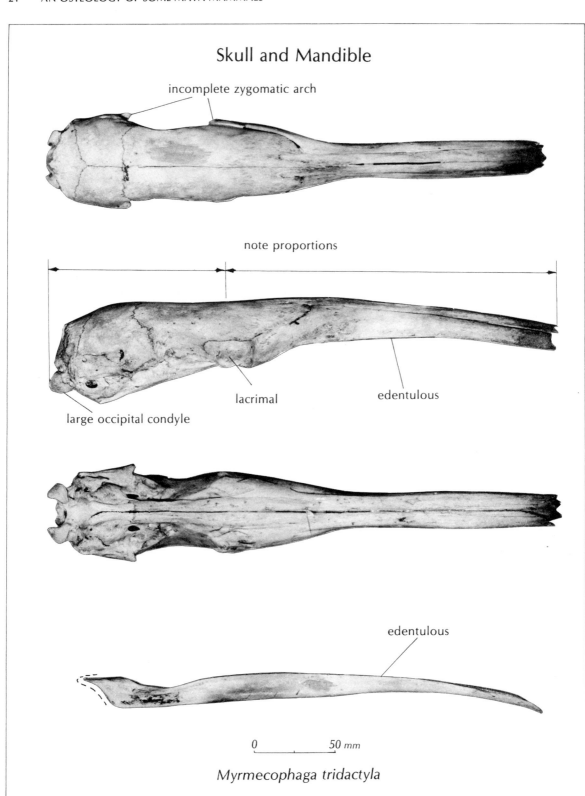

Skull and Mandible

incomplete zygomatic arch

note proportions

lacrimal

edentulous

large occipital condyle

edentulous

0   50 mm

Myrmecophaga tridactyla

FIGURE 15

Skull and Mandible

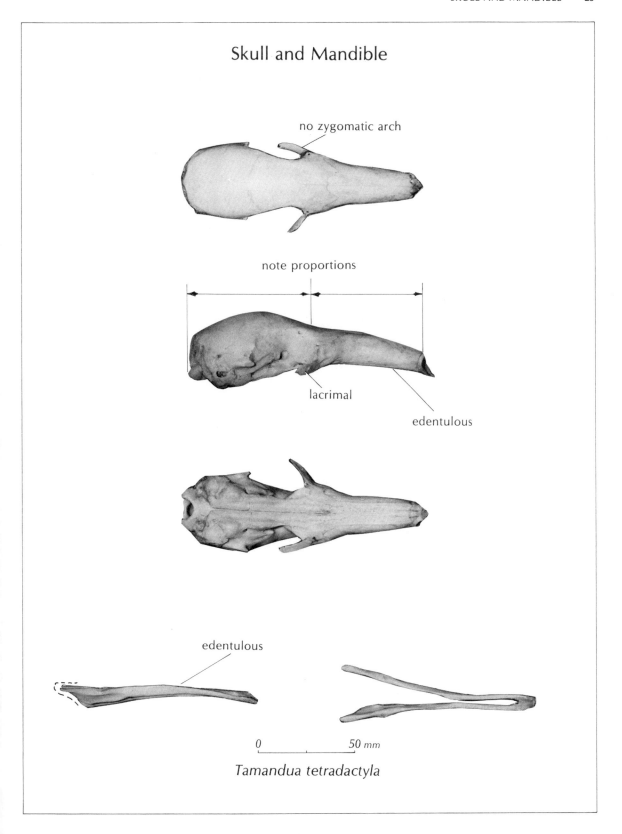

no zygomatic arch

note proportions

lacrimal

edentulous

edentulous

0   50 mm

Tamandua tetradactyla

FIGURE 16

Skull and Mandible

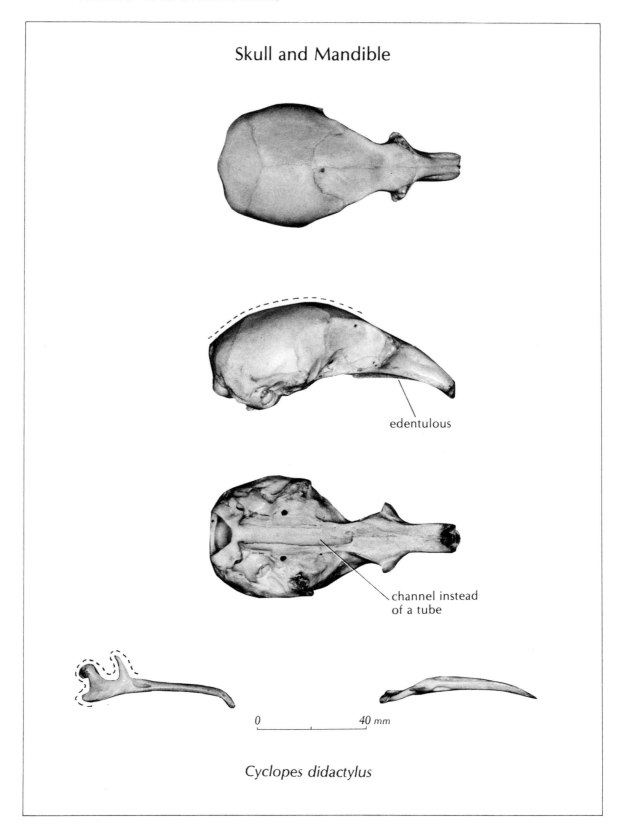

edentulous

channel instead
of a tube

0   40 mm

Cyclopes didactylus

FIGURE 17

Skull and Mandible

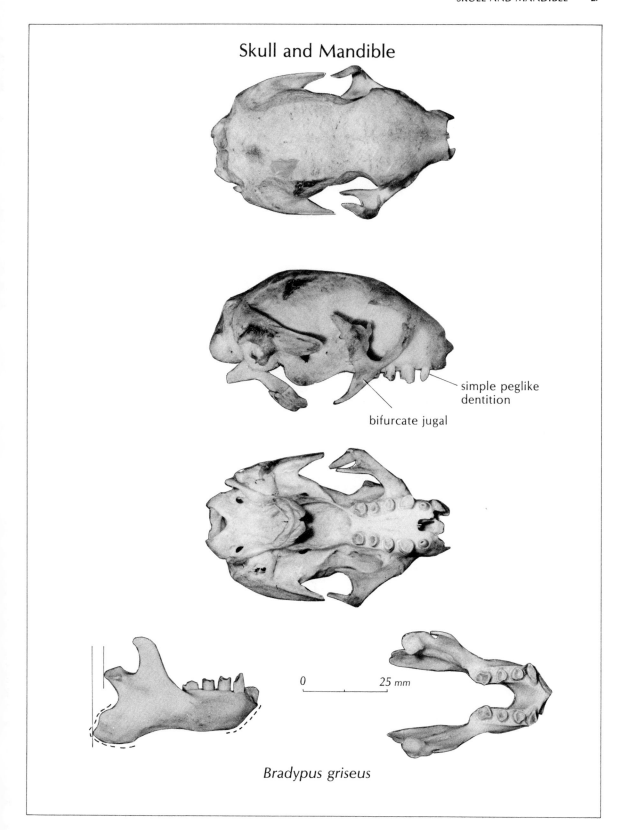

simple peglike dentition

bifurcate jugal

0   25 mm

Bradypus griseus

FIGURE 18

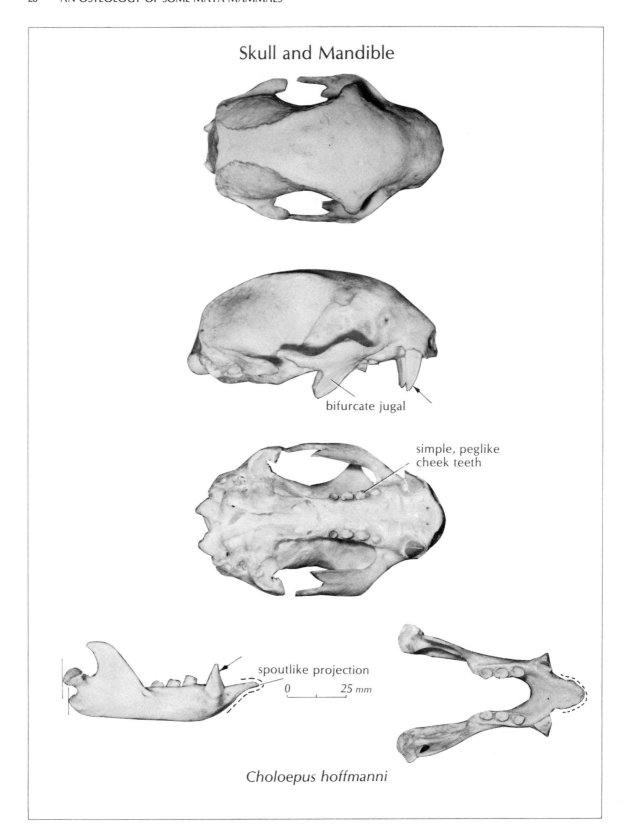

Skull and Mandible

bifurcate jugal

simple, peglike
cheek teeth

spoutlike projection

0   25 mm

Choloepus hoffmanni

FIGURE 19

Skull and Mandible

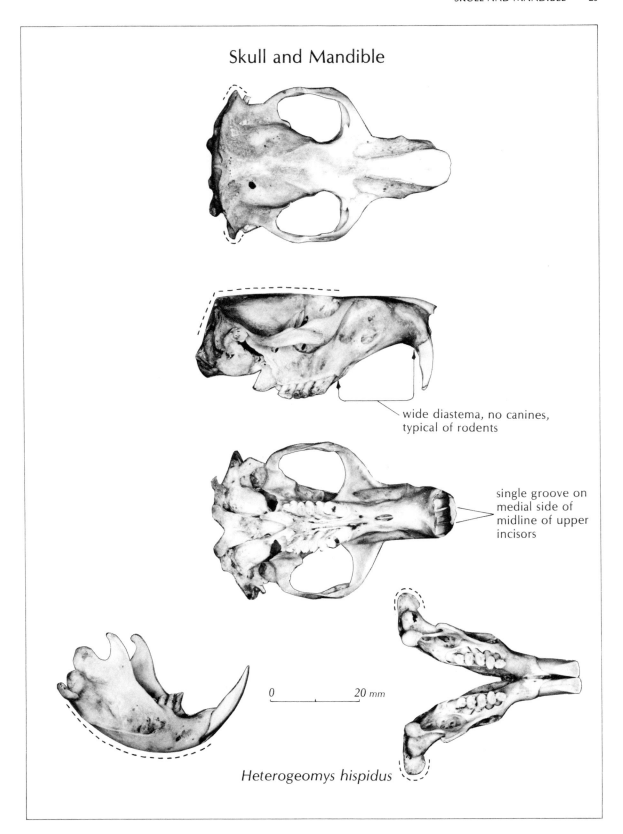

wide diastema, no canines,
typical of rodents

single groove on
medial side of
midline of upper
incisors

0   20 mm

Heterogeomys hispidus

FIGURE 20

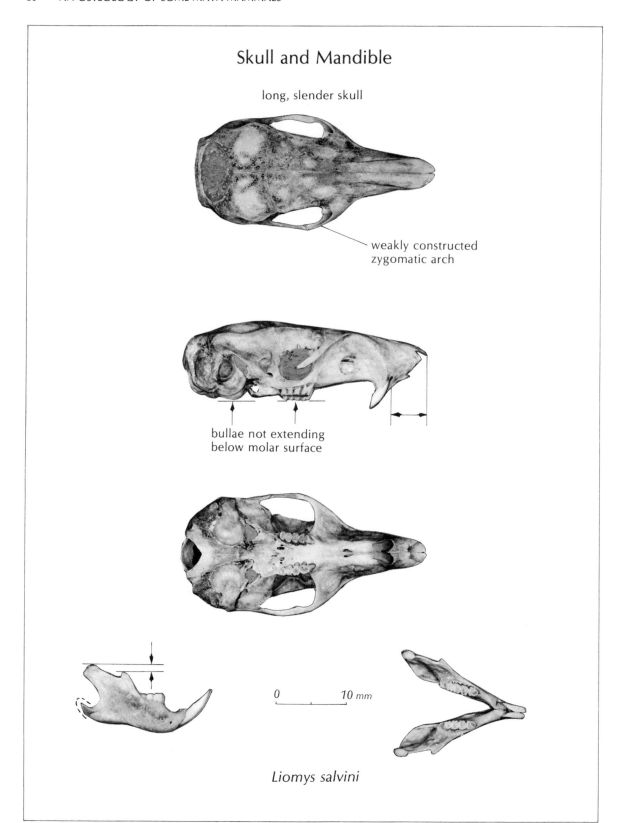

Skull and Mandible

long, slender skull

weakly constructed
zygomatic arch

bullae not extending
below molar surface

0   10 mm

Liomys salvini

FIGURE 21

Skull and Mandible

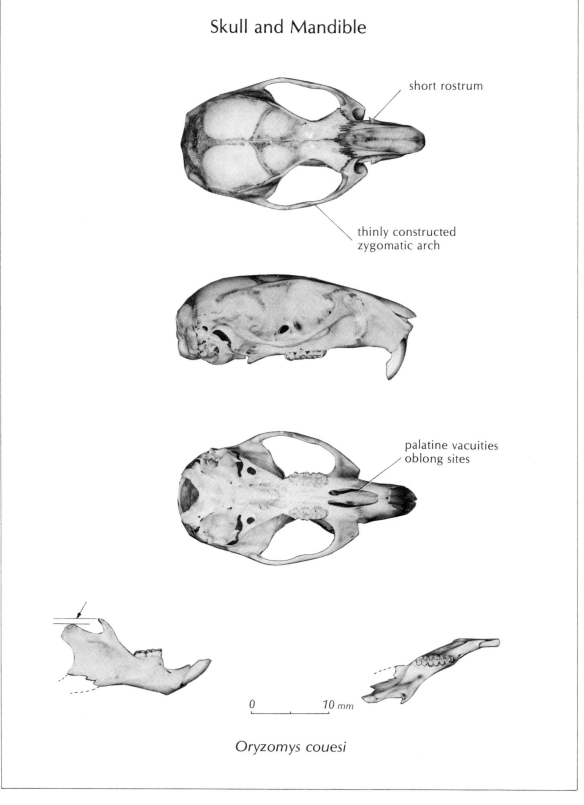

short rostrum

thinly constructed zygomatic arch

palatine vacuities oblong sites

0   10 mm

Oryzomys couesi

FIGURE 22

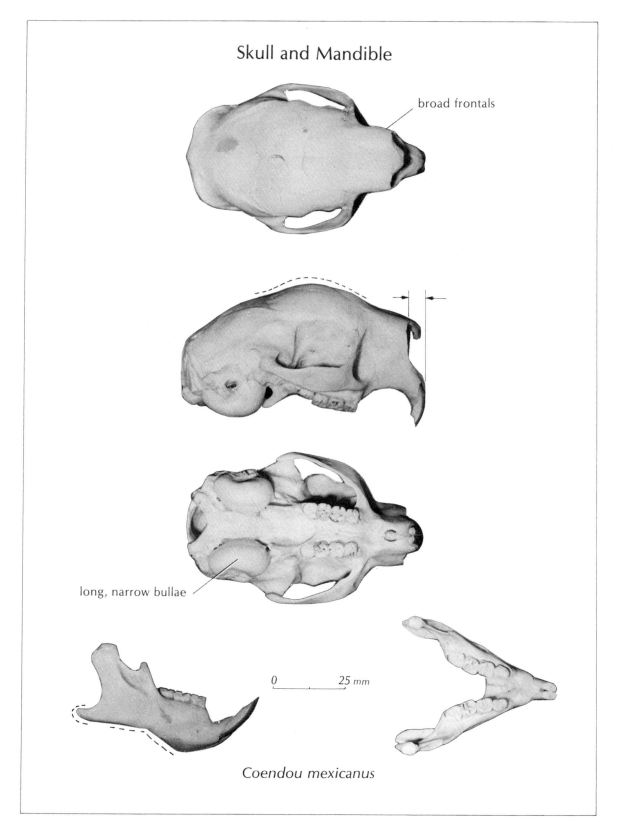

Skull and Mandible

broad frontals

long, narrow bullae

0   25 mm

Coendou mexicanus

FIGURE 23

Skull and Mandible

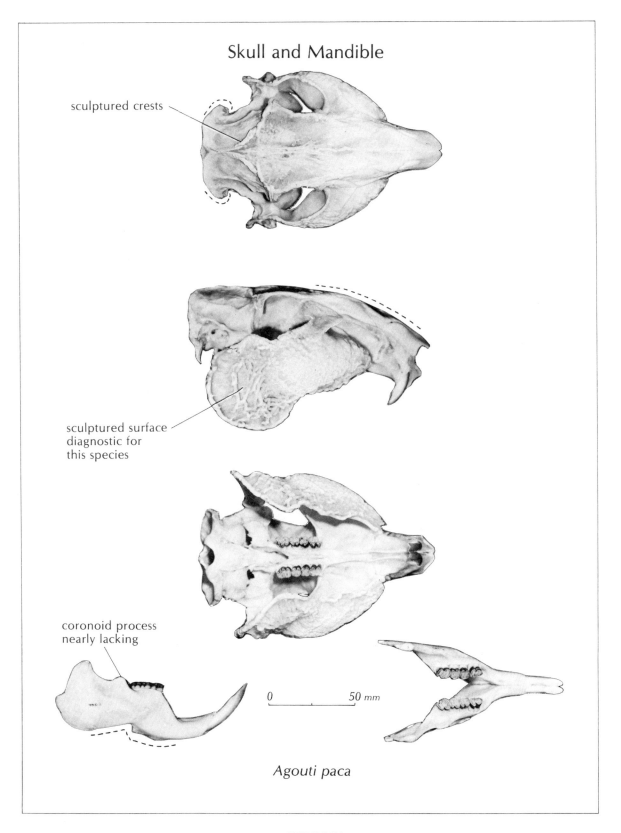

sculptured crests

sculptured surface
diagnostic for
this species

coronoid process
nearly lacking

0   50 mm

Agouti paca

FIGURE 24

Skull and Mandible

long, slender skull

no sagittal crest

prominent process

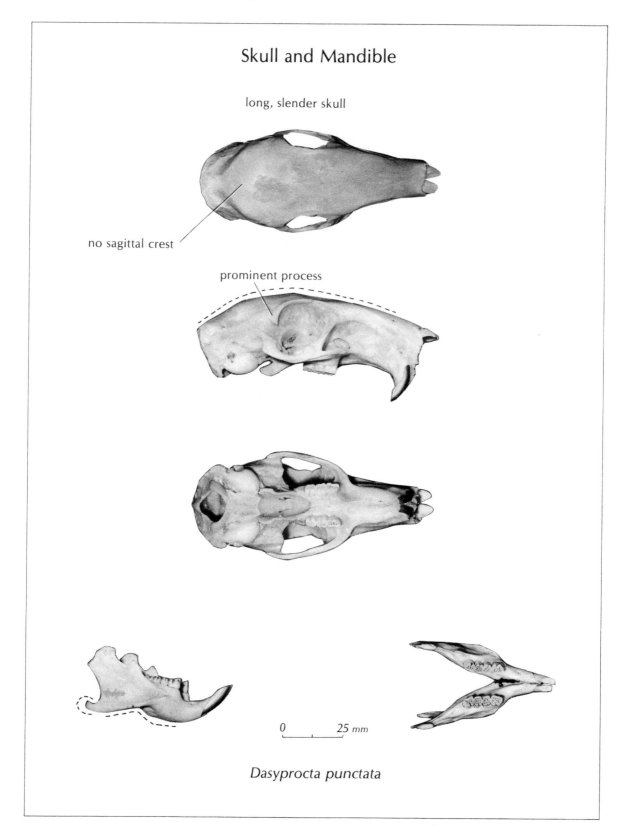

0   25 mm

Dasyprocta punctata

FIGURE 25

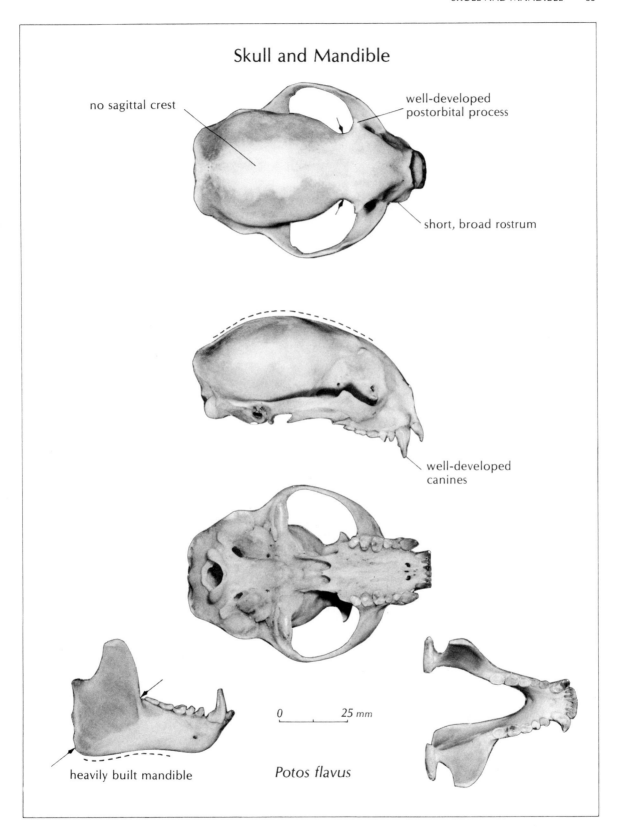

Skull and Mandible

no sagittal crest

well-developed postorbital process

short, broad rostrum

well-developed canines

0   25 mm

heavily built mandible

Potos flavus

FIGURE 26

Skull and Mandible

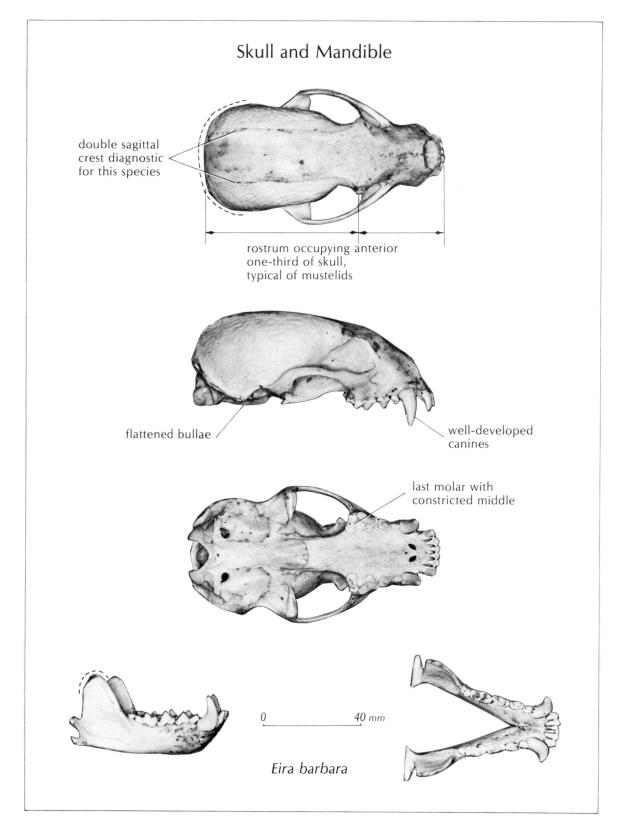

double sagittal crest diagnostic for this species

rostrum occupying anterior one-third of skull, typical of mustelids

flattened bullae

well-developed canines

last molar with constricted middle

0   40 mm

Eira barbara

FIGURE 27

Skull

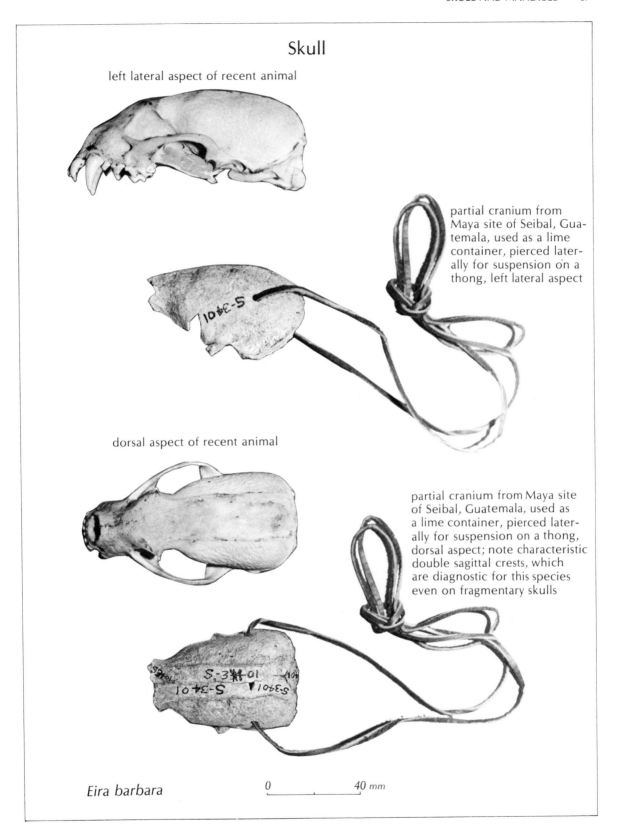

left lateral aspect of recent animal

partial cranium from Maya site of Seibal, Guatemala, used as a lime container, pierced laterally for suspension on a thong, left lateral aspect

dorsal aspect of recent animal

partial cranium from Maya site of Seibal, Guatemala, used as a lime container, pierced laterally for suspension on a thong, dorsal aspect; note characteristic double sagittal crests, which are diagnostic for this species even on fragmentary skulls

Eira barbara

0 _____ 40 mm

FIGURE 28

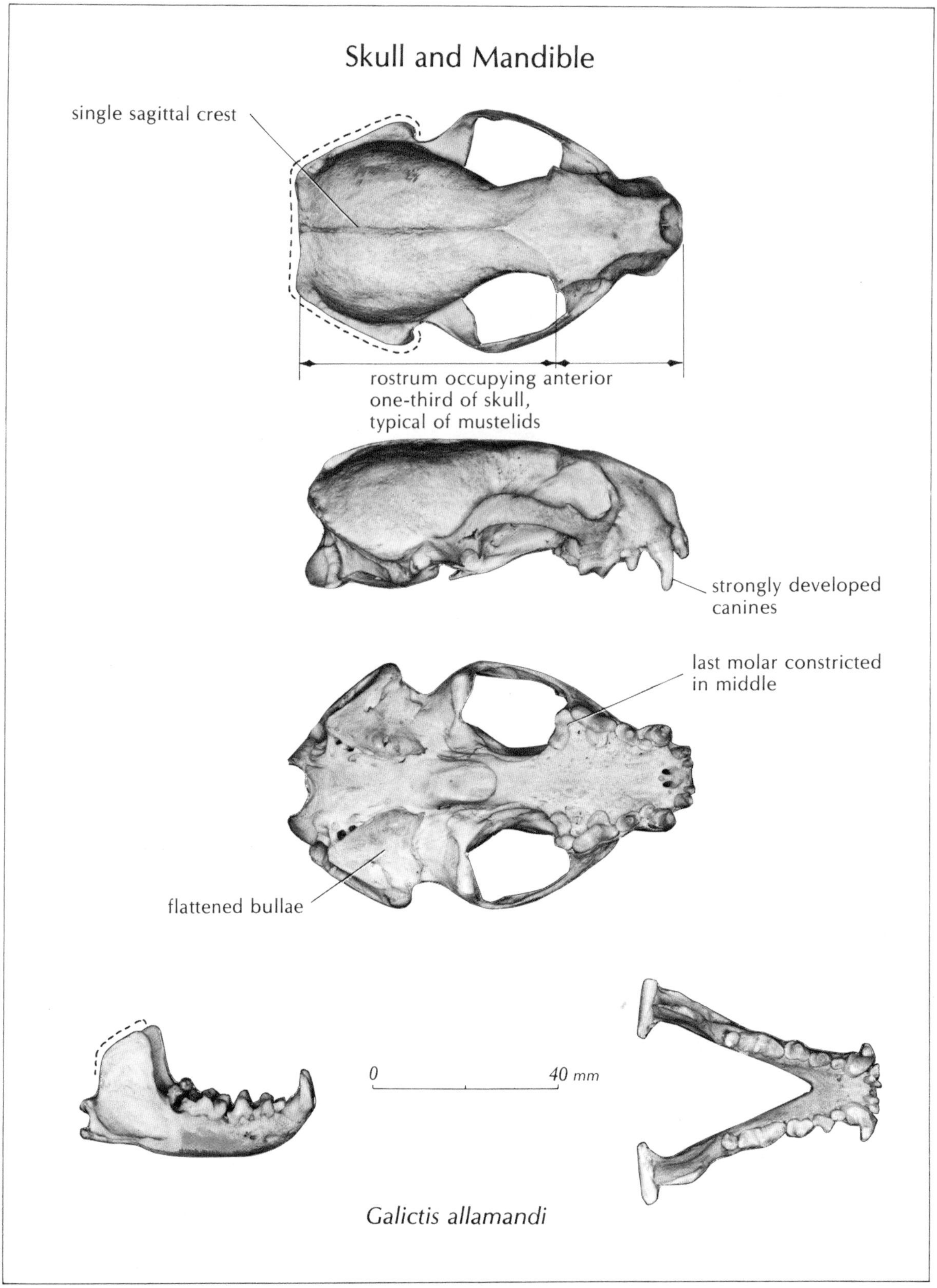

Skull and Mandible

single sagittal crest

rostrum occupying anterior
one-third of skull,
typical of mustelids

strongly developed
canines

last molar constricted
in middle

flattened bullae

0   40 mm

Galictis allamandi

FIGURE 29

Skull and Mandible

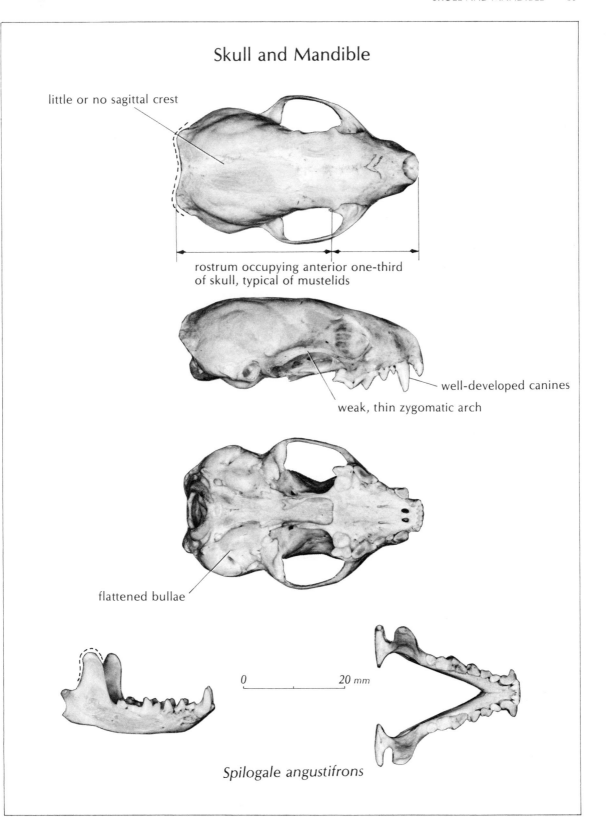

little or no sagittal crest

rostrum occupying anterior one-third of skull, typical of mustelids

well-developed canines

weak, thin zygomatic arch

flattened bullae

0   20 mm

Spilogale angustifrons

FIGURE 30

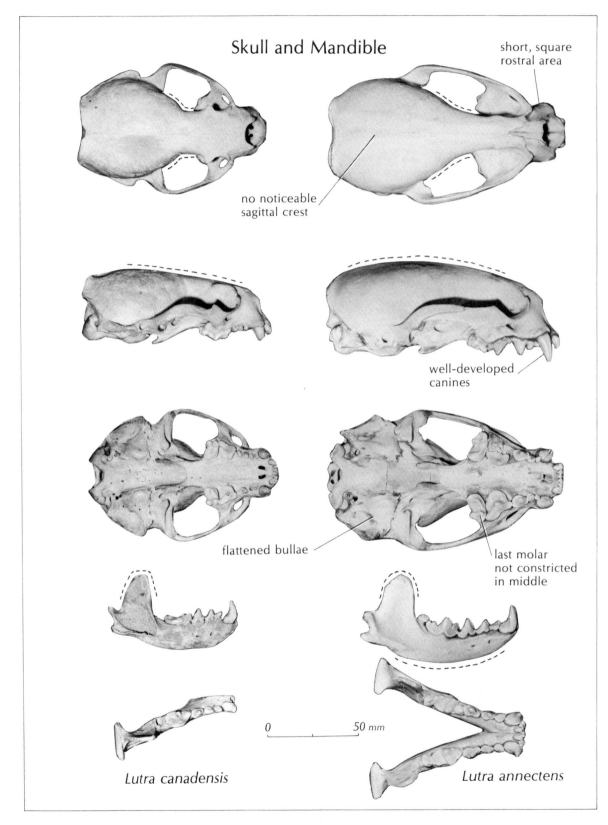

Skull and Mandible

short, square
rostral area

no noticeable
sagittal crest

well-developed
canines

flattened bullae

last molar
not constricted
in middle

0   50 mm

Lutra canadensis   *Lutra annectens*

FIGURE 31

Skull and Mandible

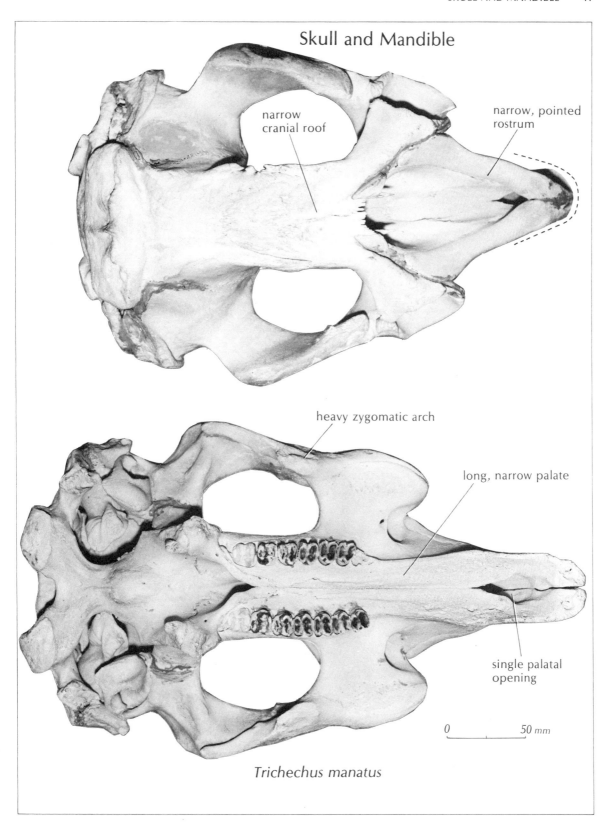

narrow
cranial roof

narrow, pointed
rostrum

heavy zygomatic arch

long, narrow palate

single palatal
opening

0 _____ 50 *mm*

Trichechus manatus

FIGURE 32

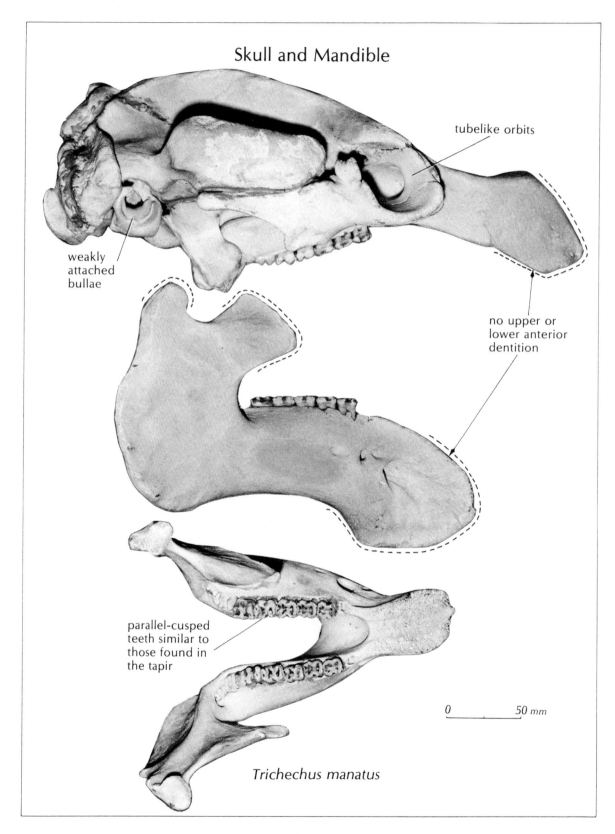

Skull and Mandible

tubelike orbits

weakly
attached
bullae

no upper or
lower anterior
dentition

parallel-cusped
teeth similar to
those found in
the tapir

0   50 mm

Trichechus manatus

FIGURE 33

Skull and Mandible

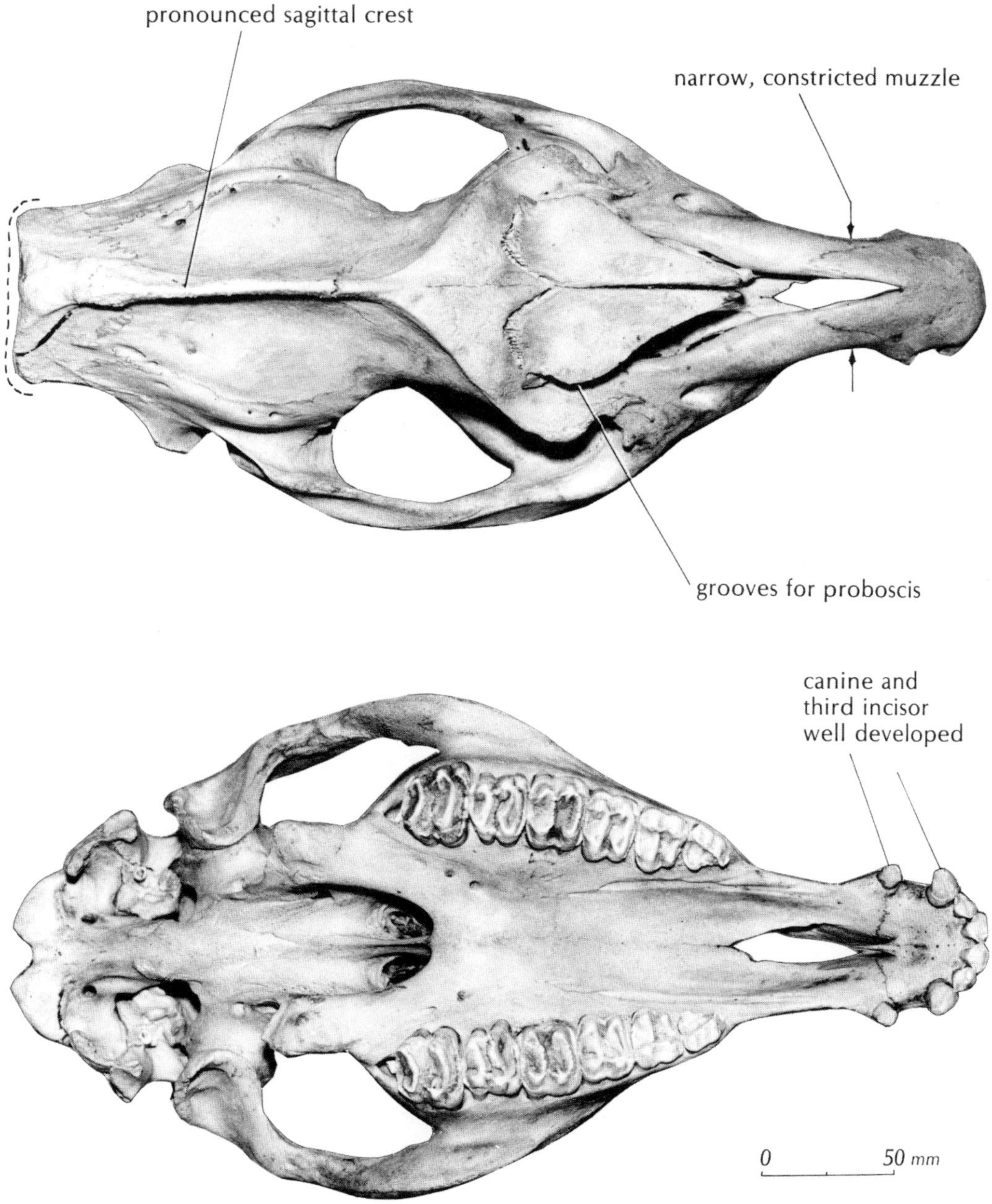

pronounced sagittal crest

narrow, constricted muzzle

grooves for proboscis

canine and
third incisor
well developed

0   50 mm

Tapirus bairdii

FIGURE 34

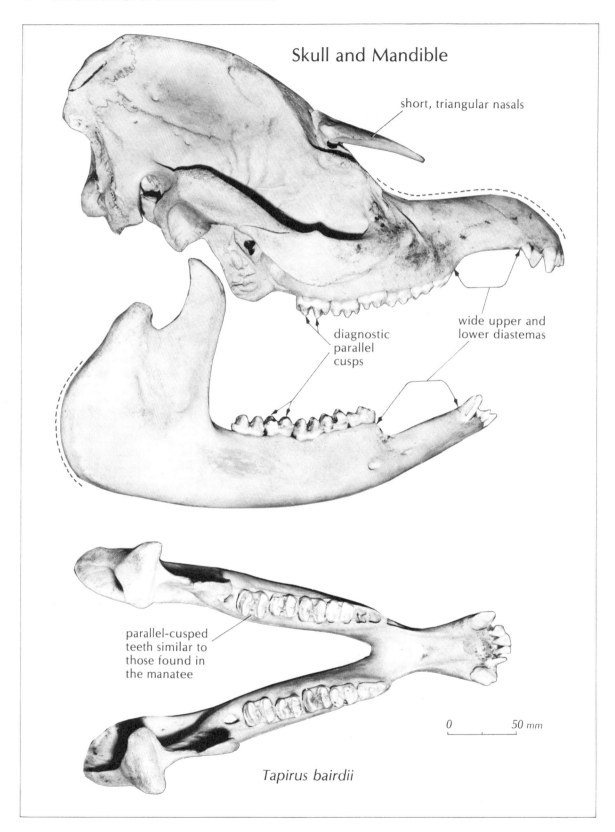

Skull and Mandible

short, triangular nasals

diagnostic parallel cusps

wide upper and lower diastemas

parallel-cusped teeth similar to those found in the manatee

0   50 mm

Tapirus bairdii

FIGURE 35

Skull and Mandible

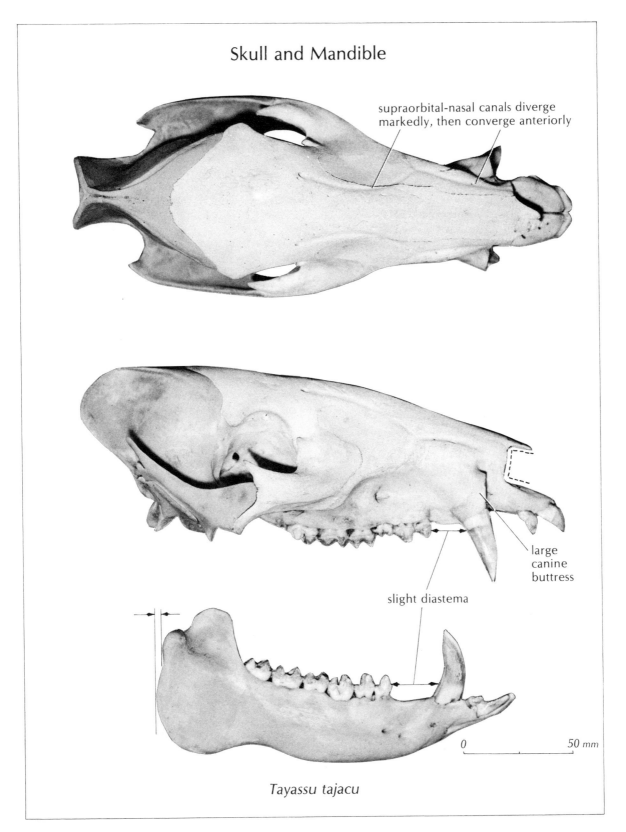

supraorbital-nasal canals diverge markedly, then converge anteriorly

large canine buttress

slight diastema

0   50 mm

Tayassu tajacu

FIGURE 36

Skull and Mandible

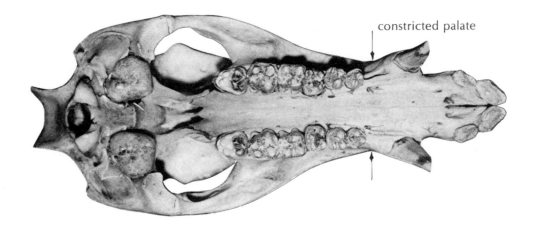

constricted palate

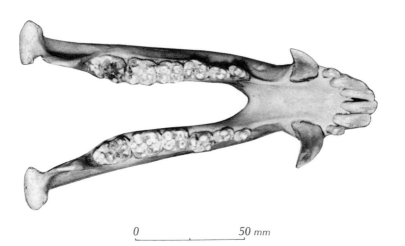

0   50 mm

Tayassu tajacu

FIGURE 37

Skull and Mandible

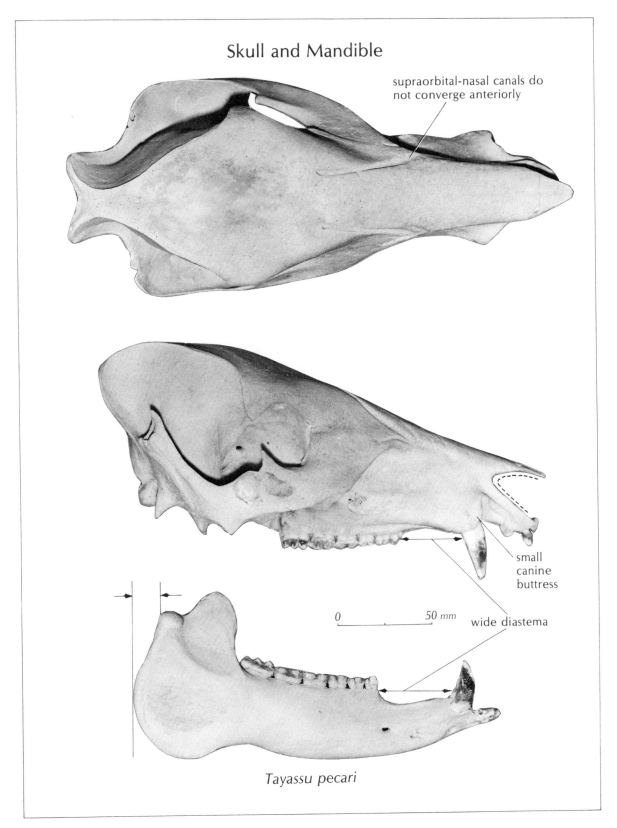

supraorbital-nasal canals do
not converge anteriorly

small
canine
buttress

wide diastema

0   50 mm

Tayassu pecari

FIGURE 38

Skull and Mandible

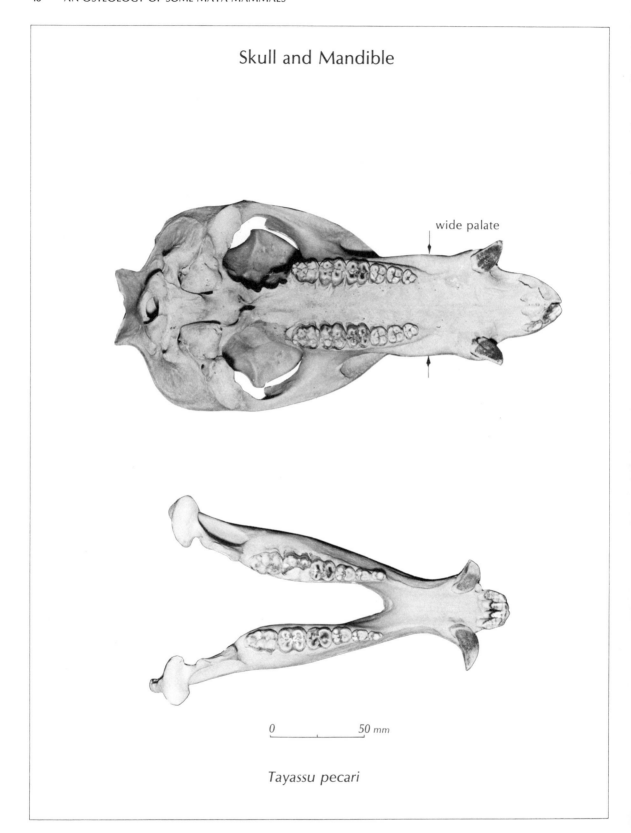

wide palate

0 ⟷ 50 mm

Tayassu pecari

FIGURE 39

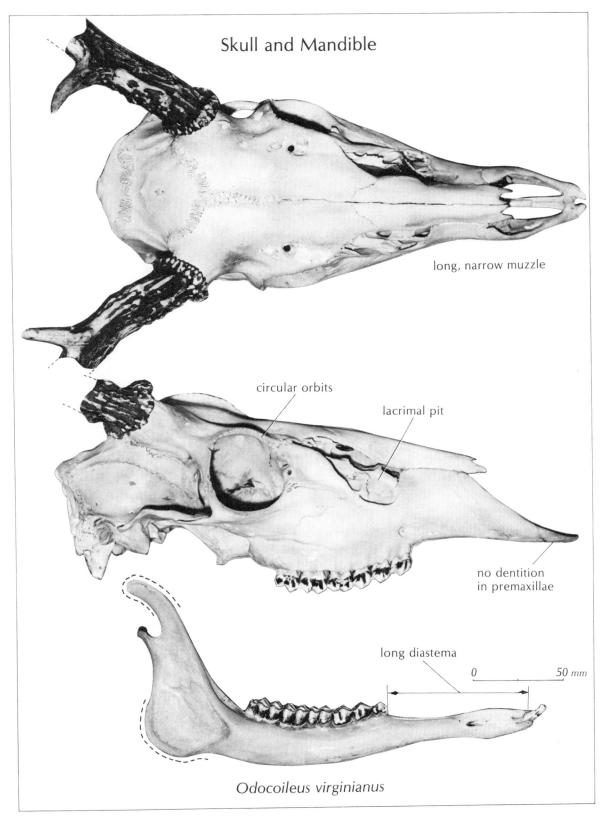

Skull and Mandible

long, narrow muzzle

circular orbits

lacrimal pit

no dentition
in premaxillae

long diastema

0   50 mm

Odocoileus virginianus

FIGURE 40

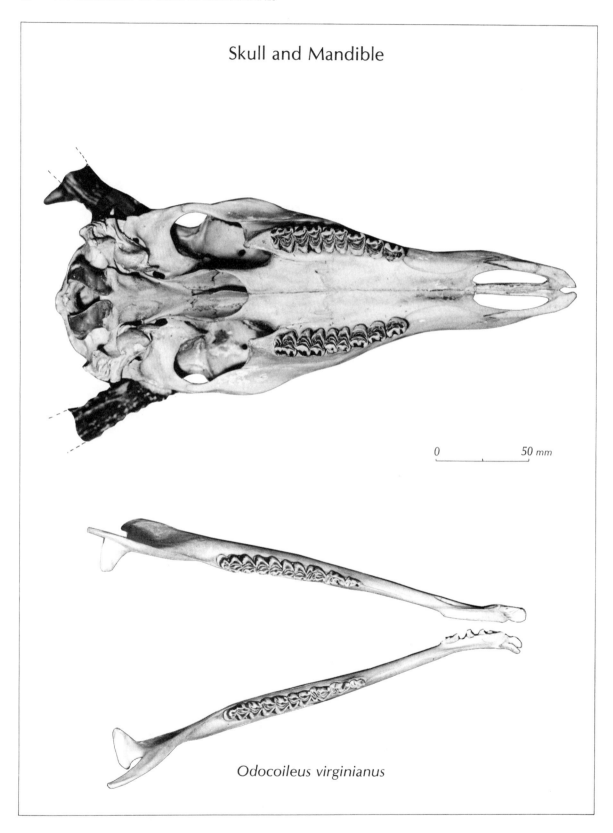

Skull and Mandible

0 ⸺⸺⸺ 50 mm

Odocoileus virginianus

FIGURE 41

Skull and Mandible

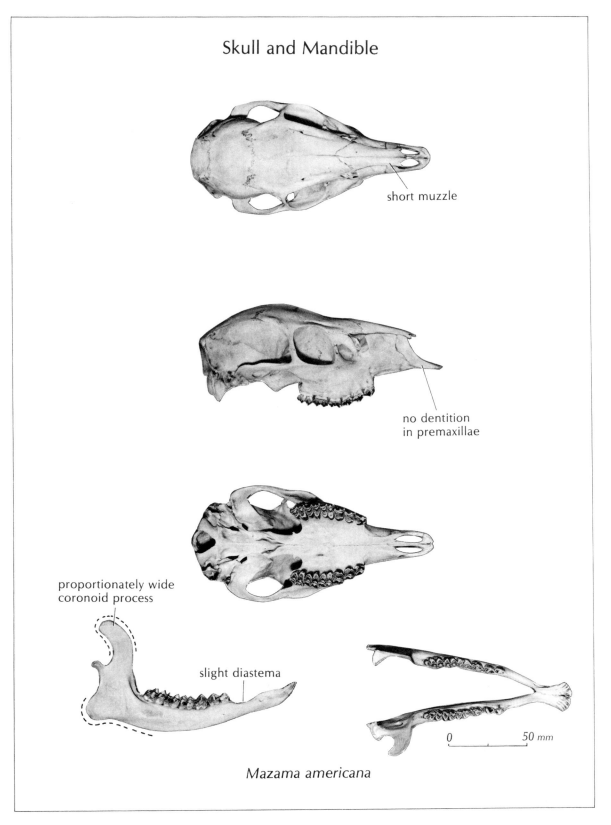

short muzzle

no dentition
in premaxillae

proportionately wide
coronoid process

slight diastema

0   50 mm

Mazama americana

FIGURE 42

Skull

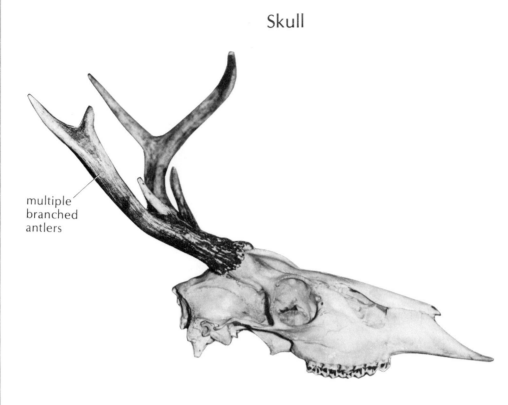

multiple
branched
antlers

Odocoileus virginianus

0 — 50 *mm*

single spike

Mazama americana

note size difference

FIGURE 43

Postcranial Skeleton

As with the skulls, the size scales must be taken into consideration when comparing elements of the postcranial skeletons. In order for the illustrations on the plates to be uniform, all examples of similar bones occupy the same area regardless of their proportional differences. Therefore, both bat and tapir humeri are represented as having the same length. The size scales will indicate the actual sizes of the various bones.

Scapula

Order: Marsupialia — Opossums *(fig. 44)*

The scapulae of all three opossums have straight axillary borders and expanded coracoid borders. The acromion process is widest and longest in *Caluromys derbianus*. The small size of *Marmosa mexicana* separates this opossum from the other two, which are of nearly equal size. The scapula of *Philander opossum* has a slightly hooked coracoid process. The spines of all three separate the blades into two nearly equal areas.

Order: Chiroptera — Bats *(fig. 44)*

In general the scapula is roughly oval and consists chiefly of a postscapular fossa, the prescapular fossa being extremely small. The former is divided into subfossae by low ridges. The scapular spine is short in length and moderately deep with a large, simple acromion process. The coracoid process is long and thin. Bat scapulae are quite similar among many families and genera, making it difficult to taxonomically distinguish species based on this element alone.

Note that the angle formed at the junction of the vertebral and axillary borders is the only readily noticeable area where the scapulae of the two widely differing species *Artibeus lituratus* and *Desmodus rotundus* can be distinguished (fig. 44). In the former the apex is rounded, while in the latter it terminates in a point. This minor variability in scapular form extends throughout the families of bats in general.

Order: Primates — Monkeys *(figs. 44, 45)*

The scapulae are rather wide in both the howler monkey, *Alouatta villosa*, and in the spider monkey, *Ateles geoffroyi*. The coracoid and axillary borders of the blades are of nearly equal length in both monkeys. There are coracoscapular foramina in both. The coracoid processes are well developed in both, but there is a greater extension beyond the blade in the spider monkey. The acromion processes are large in both forms. In *Ateles geoffroyi*, it is directed along the axis of the scapular spine, while in *Alouatta villosa*, it is directed at an angle away from the spine. The scapular spines in both primates bisect the blades into nearly equal fossae.

Order: Edentata — Anteaters and Sloths *(fig. 45, 46)*

The scapulae of anteaters in general are broad and rounded, having no distinct angle separating the vertebral and axillary borders. The coracoid borders join the coracoids, converting the coracoscapular notches into foramina, except in *Cyclopes didactylus* where the coracoscapular notch is not converted into a foramen. The blades are separated into three fossae by two spines, the main spine and an inferior scapular spine. The latter produces the postscapular fossa. The inferior spine and postscapular fossa are less developed in the two-toed anteater than in the other two. Both *Myrmecophaga tridactyla* and *Tamandua tetradactyla* have prominent processes developed midway on the surface of the main scapular spines. The acromion processes are long and slender in all three anteaters. There are no distinct metacromions in any of the three. Size alone is a distinguishing factor in separating these three similarly shaped edentates. The scapula of *Tamandua tetradactyla* is half of the size of this bone in *Myrmecophaga tridactyla*, while that of *Cyclopes didactylus* is scarcely one-fifth of the size of this bone in the giant anteater.

The scapulae of the sloths are irregular in form, having high, rounded apexes at the union of the vertebral and axillary borders. Both scapulae in *Bradypus griseus* and *Choloepus hoffmanni* have coracoscapular foramina. The prescapular fossae are larger than the postscapular fossae in sloths. The spines in both sloths arise from midpoints on the scapulae, but in the two-toed sloth, the spine extends into a long acromion process, curving to almost unite with the coracoid. The acromion process is reduced in size in *Bradypus griseus*, not approaching a connection with the coracoid. The vertebral border in *Choloepus hoffmanni* is noticeably notched.

Order: Rodentia — Rodents *(fig. 46)*

Only five examples have been chosen to represent the myriads of rodents that inhabit the Maya area. The two smaller forms, *Heterogeomys hispidus* and *Liomys salvini*, have scapulae that are typical of most small rodents. The scapulae of the three larger animals, *Coendou mexicanus*, *Agouti paca*, and *Dasyprocta punctata*, are representative of the larger rodents in general.

Typically rodent scapulae are high and narrow. The acromion processes are long and prominent. The acromioscapular notches are deep to the point that the actual scapular spines occupy only short areas near the suprascapular borders. The coracoid processes are generally rather small, blunt, and at times hooked.

In *Heterogeomys hispidus*, the vertebral border is quite wide with a decided overhang at the angle produced by the union of the vertebral and axillary borders. The presence of an inferior scapular spine forms a substantial postscapular fossa. The coracoid process is prominent, having a decided hook at the end. The acromion process is thin and extended. The scapula of *Liomys salvini* has a noticeable outward bow or swelling of the coracoid border. The spine is located midway on the blade. There is no coracoid process. The rather long acromion process terminates in a round, flattened end. The Mexican porcupine, *Coendou mexicanus*, has a scapula with a wide, rounded blade. The spine is high and thin with an acromion process that has a bifurcate terminus. There is also a decided coracoid process. The scapula of *Agouti paca* is paddle-shaped in form. The spine, which is located centrally on the blade, is stout and prominent. The acromion process terminates in a flat, triangular end. The scapula of *Dasyprocta punctata* is rather wide and rounded on the coracoid border, the axillary border being straight. The spine is stout and heavy and terminates in a wide, flat end, which is set off to the axillary side of the scapular spine.

Order: Carnivora — Kinkajous and Mustelids *(figs. 46, 47)*

The kinkajou *(Potos flavus)* scapula is rhomboidal, approximating that of other procyonids and some ursids. The spine is located obliquely across the blade. A slight inferior scapular spine and accompanying postscapular fossa are also present. The acromion process is well developed and directed at an angle toward the scapular notch.

The mustelids, *Eira barbara*, *Galictis allamandi*, and *Lutra annectens*, have scapulae that differ from each other in several aspects. *Eira barbara* has a scapula that has a rounded coracoid border; the axillary border is nearly straight. The acromion process has a wide, flar-

ing crest that diminishes into a thin terminal end. The glenoid is strongly constructed. The scapula of *Galictis allamandi* has a rectangular blade with only a mere suggestion of a postscapular fossa. The spine is relatively prominent, terminating in a moderate acromion process. The scapula of *Lutra annectens* is extremely bowed or rounded on the coracoid border. The axillary border is straight with a decided overhang at the union of this border with the vertebral border. The spine is strong and moderately high. The acromion process has a rounded prominence directed toward the axillary border.

Order: Sirenia — Manatees *(fig. 47)*

The manatee, *Trichechus manatus*, has a scapula (opposite side from other views) with a moderately wide blade. The axillary border is straight. The coracoid border is strongly curved or bowed. The spine arises from about midway on the surface of the blade. There is a noticeable swelling at its base. The acromion process is long and thin and directed forward toward the coracoid border. The glenoid is curved toward the coracoid border.

Order: Perissodactyla — Tapirs *(fig. 47)*

The scapula (opposite side from other views) of Baird's tapir, *Tapirus bairdii*, is rather long and narrow with a rounded vertebral border. The axillary border is straight. The coracoid border is curved or bowed with a decided notch at the neck. The spine is short and peaked with a widened process at its midpoint. There is a decided overhang at the angle formed by the union of the vertebral and axillary borders. There is a slightly hooked coracoid process. There is no acromion process.

Order: Artiodactyla — Peccaries and Deer *(figs. 47, 48)*

The scapulae of the four artiodactyls figured and discussed here can be collectively separated from the other forms in this study by their relatively large size (except *Mazama americana*) and by the shape of their blades, which is an inverted isosceles triangle. In all four animals the vertebral borders are straight. The necks are rather long and considerably constricted. All have rather prominent spines running most of the length of the blades near the coracoid borders. The spine ends considerably short of the neck in *Tayassu tajacu* with no acromion process or noticeable coracoid process. Also in *Tayassu tajacu*, there is a vertical arching of the spine on its upper half, curving slightly toward the axillary border.

In *Tayassu pecari* the blade is almost like that of the

other peccary, except the process on the spine is longer and not as developed. *Odocoileus virginianus* has the widest blade of the four. The spine is quite high and thickened at the midpoint of the crest, remaining in line with the spine proper. There is a slight acromion process and only a trace of a coracoid swelling. The comparatively small size of the scapula of *Mazama americana* is generally adequate in separating it from the other three similarly shaped blades. The spine is quite prominent and thin with no thickening of its upper or outer edge. A rather well-defined acromion process and a coracoid process are also present. The neck is noticeably thinner than that found in *Odocoileus*.

Humerus

Order: Marsupialia — Opossums *(fig. 48)*

The humerus of *Philander opossum* is the largest of the three animals being described. It has moderate tuberosities hardly projecting above the articular head. The deltoid ridge is low. The supinator ridge is well developed and low. The medial epicondyle is large, as is the entepicondylar foramen. In *Marmosa mexicana*, the lateral and medial tuberosities are also slight. The medial epicondyle is quite prominent. A moderate entepicondylar foramen is present. There is no defined supinator ridge or defined deltoid ridge. In *Caluromys derbianus*, the medial tuberosity is stout but not projecting above the head. The humerus of this opossum has the greatest development of the supinator ridge. There is a decided notch at its union with the shaft of the humerus. The medial epicondyle, as in the other species, is prominent. The entepicondylar foramen is well developed.

Order: Chiroptera — Bats *(fig. 49)*

In general the humeri of bats are long and slender, and have a slight curve. The only crest or ridge present on the bat humerus is the pectoral ridge, which lies on the proximal end of the bone starting at a point between the tuberosities and ending a short way down the neck. The spinous process on the distal trochlea and the articular trochlea itself are comparatively large and strongly developed. There is no entepicondylar foramen. The humeri of both *Artibeus lituratus* and *Desmodus rotundus*, as well as of other chiropters, have minute distinguishing characteristics that cannot be readily defined without a great many identified comparative skeletons on hand. Generally with this group of mammals, a specialist who has worked with the variations within this order is required to differentiate among the various bones of the families and genera under scrutiny.

Order: Primates — Monkeys *(fig. 49)*

Both monkeys, *Alouatta villosa* and *Ateles geoffroyi*, have humeri that are quite long with slender shafts. The humerus of *Ateles geoffroyi* is the longer of the two. The heads in both are almost ball-like with virtually little development of the medial and lateral tuberosities. The medial tuberosities in both are the most extended. There are no entepicondylar foramina. The deltoid crests in both are less than moderate, being instead mere ridges.

Order: Edentata — Anteaters and Sloths *(figs. 49, 50)*

The entire form of the humeri of anteaters is unique to this family of edentates and is generally adequate for identification with a minimum of comparison. The medial and lateral tuberosities are low, knobby processes in both *Myrmecophaga tridactyla* and *Tamandua tetradactyla*. The distal ends of both are considerably flattened. The deltoid tuberosities on the midshafts are quite prominent. The entepicondylar foramina are almost in line with the axes of the shafts, producing canals rather than foramina. Wide, shallow grooves run from these openings the full length of the shafts on the medial faces.

The humerus of *Tamandua tetradactyla* is one-half of the size of that in *Myrmecophaga tridactyla*. The smallest of the anteaters, *Cyclopes didactylus*, is about one-fourth of the size of *Tamandua tetradactyla*. In *Tamandua tetradactyla* the flattened distal surface occupies one-half of the surface of the bone. The entepicondylar foramen is not developed as a canal. There is no continuing groove from this foramen. The deltoid tuberosity is joined to the supinator ridge to form a prominent foramen.

The humeri of both sloths have shafts that are long and thin. The distal end of the humerus of *Bradypus griseus* is widened. There is no entepicondylar foramen or supinator ridge. The distal end of this element in *Choloepus hoffmanni* is widely flattened and has a strong supinator ridge and a large entepicondylar foramen. There is no deltoid tuberosity, and the deltoid ridge is quite reduced in *Choloepus hoffmanni* and nearly absent in *Bradypus griseus*. The articular heads are prominent and quite round. The medial and lateral tuberosities are low, rounded processes.

Order: Rodentia — Rodents *(figs. 50, 51)*

In general the shafts of the humeri are long, slender, and straight. In *Heterogeomys hispidus*, the medial and lateral tuberosities are low and widespread. There is a well-defined deltoid tuberosity. The medial epicondyle is projected, but there is no entepicondylar foramen. The lateral epicondyle is moderately developed. The supratrochlear fossa is well defined. In

Liomys salvini, there is little or no development to the medial and lateral epicondyles and tuberosities. The deltoid tuberosity is present as a low crest. In *Coendou mexicanus,* the humeral shaft is comparatively heavy and stout. The medial and lateral tuberosities are low, rounded processes. The medial epicondyle is extended laterally but without an accompanying entepicondylar foramen. The supratrochlear fossa is deep. The deltoid crest is well defined and heavy. In the *Agouti paca* humerus, the medial tuberosity is strongly developed and extends above the articular head. The lateral tuberosity is present as a slight knob-like process. The deltoid ridge is prominent and extends down two-thirds of the length of the shaft. There is no entepicondylar foramen, but there is a large supratrochlear perforation. In *Dasyprocta punctata,* the medial tuberosity is slightly larger than the lateral process, but neither is developed to the degree of that found in *Agouti paca.* The deltoid ridge is barely noticeable. There is no entepicondylar foramen. There is a large supratrochlear foramen.

Order: Carnivora — Kinkajous, Tayras, and Grisons *(fig. 51)*

The humerus in *Potos flavus* has proximal tuberosities not developed beyond mere knobs. There is a strong supinator ridge. The entepicondylar foramen is large. There is little or no deltoid ridge. The humeral shaft of *Eira barbara* is relatively long and thin. The medial and lateral tuberosities are slightly produced. The supinator ridge is moderate. The entepicondylar foramen is prominent. This same element in *Galictis allamandi* has a large, strong articular head with a large medial tuberosity projecting to the side rather than above the head. The deltoid ridge is moderately developed and lies high on the shaft. The supinator ridge is rather short but high. The entepicondylar foramen is quite large.

Order: Sirenia — Manatees *(fig. 51)*

The humerus of the manatee, *Trichechus manatus,* is representative of one of the two largest mammals in the area, the other being the tapir. It is a heavy, stoutly constructed bone. The articular head is quite large, round, and prominent. The two proximal tuberosities are heavy and low. The deltoid ridge occupies the upper half of the element. There is no supinator ridge or entepicondylar foramen.

Order: Perissodactyla — Tapirs *(fig. 52)*

The humerus of *Tapirus bairdii* has well-developed proximal tuberosities, medial and lateral, of about equal size. The olecranon fossa is quite deep. A prominent deltoid tuberosity and a teres tuberosity are present. These tuberosities plus the large size will allow for identification of the tapir humerus.

Order: Artiodactyla — Peccaries and Deer *(fig. 52)*

The peccary humeri are generally so similar in most respects that it is difficult if not impossible to separate these elements of *Tayassu tajacu* from those of *Tayassu pecari.* The medial tuberosities extend higher above the head than they do in the cervids. The medial epicondyles also project farther beyond the articular trochleas in the peccaries than they do in the deer. The supratrochlear perforations are quite prominent in both peccaries. This area is a closed fossa in the deer. There are no entepicondylar foramina or supinator ridges in these artiodactyls. The small size of the humerus of *Mazama americana* will distinguish it from that of *Odocoileus virginianus.*

Radius and Ulna

Order: Marsupialia — Opossums *(fig. 53)*

The radii and ulnae of the three opossums are separate bones with well-developed condyles and processes. The olecranon processes are moderately developed. The radii and ulnae of all three marsupials are considerably flattened in cross section but are particularly so in *Caluromys derbianus.*

Order: Chiroptera — Bats *(fig. 53)*

In most bats the radii are extemely long and slender. The ulnae are reduced to mere vestiges and are fused to the radii about midway on their shafts. The proximal ends of the ulnae are devoid of defined processes that are common to this element in most mammals. Each bat radius has a decided hook on the lip of the proximal articular surface. These elements are not diagnostic in identifying and separating bats in general.

Order: Primates — Monkeys *(figs. 53, 54)*

In general the radii and ulnae of monkeys are rather long and slim. The radii are curved; the ulnae are straighter. The olecranon processes are not very developed. In both *Alouatta villosa* and *Ateles geoffroyi,* the distal ends of the radii are noticeably expanded.

Order: Edentata — Anteaters and Sloths *(fig. 54)*

Myrmecophaga tridactyla has a wide, thin ulna with a grooved shaft. The distal end is the same size as the proximal end. The olecranon process is well developed. The radius also has a grooved shaft. Its distal end is noticeably wider than the proximal end. The surfaces of both bones are considerably sculptured. The ulna in *Tamandua tetradactyla* is also wide but has a noticeably smoother surface than that in *Myrmecophaga tridactyla*. The radius is wider at its distal end. The shaft is decidedly grooved. In both *Myrmecophaga tridactyla* and *Tamandua tetradactyla*, each olecranon process has a decided overhang on the back of the crest. In *Cyclopes didactylus*, the small size of the elements alone will separate this species from the other anteaters. Both the radius and ulna are flattened. The distal ends of both bones are wider than their proximal ends. The olecranon crest also overhangs as in the two larger anteaters. The radii and ulnae in the sloths, *Bradypus griseus* and *Choloepus hoffmanni*, are long and slender. The radii have widely expanded distal ends, that of *Choloepus hoffmanni* being noticeably flatter on the distal one-half. The semilunar notches of both sloths are quite open and shallow. The olecranon processes of both sloths are quite low.

Order: Rodentia — Rodents (fig. 55)

The radii and ulnae of rodents are similar in many respects. They are separate bones, the latter being stouter than the radii. In *Heterogeomys hispidus*, the ulna is quite wide and flattened with a slightly grooved surface. The olecranon is noticeably high. The distal end of the radius is quite wide. In *Liomys salvini*, both elements are thin and long. The olecranon process is moderate. In *Coendou mexicanus*, the ulna is quite flattened. The olecranon process is quite low. The radius is heavier than the ulna and considerably compressed, with a distal extremity that is wider than the proximal end. Both the radius and ulna are heavily constructed in *Agouti paca*. The olecranon process is moderately strong and bluntly developed. The shaft of the ulna is deeply grooved. The semilunar notch is wide and strong. The radius is considerably curved with a well-developed head and a widened distal end. In *Dasyprocta punctata*, the ulna is heavier than the radius with a prominent groove running the length of the shaft. The olecranon process is moderately extended. The radius has well-developed proximal and distal condyles of about equal size.

Order: Carnivora — Procyonids and Mustelids (figs. 55, 56)

The radius and ulna in *Potos flavus* are of about equal size. The proximal and distal condyles are of equal size, but the olecranon process has a decided curve causing it to overhang the articulation of the semilunar notch with the articular head of the radius. The radius and the ulna in *Eira barbara* are of about equal weight. The olecranon process is quite blunt and heavily constructed. The styloid process is strongly developed and extended with a noticeable notch on the terminal surface. The radius is rather wide distally. The radius of *Galictis allamandi* is considerably twisted with a wide grooving and expansion of its distal half. The shaft of the ulna is quite flattened with a well-developed proximal end and olecranon process. Both the radius and ulna of *Lutra annectens* are stoutly constructed with relatively heavy, enlarged terminal ends. The distal ends of both bones are blunt. The olecranon process is wide and strongly developed.

Order: Sirenia — Manatees *(fig. 56)*

Movement in the front limbs of the manatee, *Trichechus manatus*, is somewhat restricted. This is reflected in the radius and ulna, which are heavily constructed, large bones that are fused together at both their proximal and distal ends. The radius is quite curved. There is little or no overhang of the semilunar notch.

Order: Perissodactyla — Tapirs *(fig. 56)*

The radius and ulna of *Tapirus bairdii* are fused at their distal ends. The olecranon process is rather highly extended and wide. The radius is heavy with a well-developed articular, proximal end.

Order: Artiodactyla — Peccaries and Deer *(figs. 56, 57)*

The radii and ulnae of both *Tayassu tajacu* and *Tayassu pecari* are fused together at their proximal ends. The olecranon process is generally a bit more extended in *Tayassu tajacu* than in *Tayassu pecari*. The olecranon processes of both *Odocoileus virginianus* and *Mazama americana* are proportionately smaller than they are in the peccaries. The small size of *Mazama americana* will separate this form from the other artiodactyls. The radius and ulna may fuse to varying degrees, or they may be separate bones.

Pelvis

Order: Marsupialia — Opossums *(fig. 57)*

The innominates of all three opossums, *Philander opossum*, *Marmosa mexicana*, and *Caluromys derbianus*, are all quite similar in form, but *Marmosa mexicana* is considerably smaller than the others of this group. The ilium is a simple, straight, narrow element.

The pubis and ischium are well developed around a comparatively large obturator foramen. In *Philander opossum*, the ischial tuberosity is angular, but it is a rounded border in the other two opossums. In each of the three opossums, there is a marsupial bone loosely attached to the pubis in the area of the pubic spine. However, it is highly unlikely that this bone would ever be found in articulation with the innominate (it never has been), as the soft, attaching tissue would disintegrate soon after the death of the animal, allowing the marsupial bone to drift away from the larger element.

Order: Chiroptera — Bats *(fig. 58)*

The hind legs and pelves of bats are peculiarly developed and weakly constructed because of the chiropteran habit of hanging by the hind legs. The ilium is rodlike with an asymmetrical flaring crest in *Artibeus lituratus*. This bone in *Desmodus rotundus* has a crest with projections that flare equally. The obturator foramina are proportionately large in both bats. The ischial tuberosities of both are quite pronounced as are the large pubic spines that rise above the body of the pubis. The ischial tuberosity is a rounded process in *Artibeus lituratus* and is a flattened platelike area of the ischium in *Desmodus rotundus*. In general the fragile nature of these elements in bats does not allow for their common recovery from archaeological sites.

Order: Primates — Monkeys *(fig. 58)*

The pelves in both *Alouatta villosa* and *Ateles geoffroyi* are quite similar, that of *Ateles geoffroyi* being a bit larger. The ilia are long and moderately wide. The anterior margin of the ilium is straight in *Ateles geoffroyi* and slightly concave or dished in *Alouatta villosa*. The acetabula are deep in both monkeys. The ischial tuberosities are strongly developed in both. The pubic spine is quite pointed in *Alouatta villosa* and forms more of a right angle in *Ateles geoffroyi*.

Order: Edentata — Anteaters and Sloths *(figs. 58, 59)*

The large size of the pelvis of *Myrmecophaga tridactyla* sets it off from the other, smaller anteaters. The ilium in this large species has a decided notch at its base where it joins the acetabulum. The obturator foramen is large and oval, and the border formed by the ischium and pubis is quite rounded. The acetabulum is deep. In *Tamandua tetradactyla*, both the left and right innominates tend to fuse early to the large sacrum to form a strong unit. As with the larger species, the lower part of the ilium is notched. The ischial tuberosity and the pubic spine are quite angular. The anterior iliac spine has a decidedly hooked terminus.

The smallest anteater, *Cyclopes didactylus*, also has a pelvis consisting of a firmly fused unit consisting of a left and right innominate and the sacrum. It, too, is notched at the base of the ilium where it joins a deep acetabulum. The obturator foramen is shaped irregularly rather than with the usual circular or oval opening. The small size and typical anteater form of the pelvis, just described, will usually identify this animal. The pelves of the sloths, *Bradypus griseus* and *Choloepus hoffmanni*, are unique in form. In both animals the innominates and the large, wide sacra fuse to form what appear to be single bones rather than units composed of three bones. The pubis is pointed in *Bradypus griseus* and blunted in *Choloepus hoffmanni*. These processes project from the left and right innominates to form thin, joined rings in both sloths. The ilia spread out to join the sacra as do the ischial tuberosities. The free end of the ilium is flared in *Bradypus griseus* but straighter in *Choloepus hoffmanni*.

Order: Rodentia — Rodents *(figs. 59, 60)*

The ilium of *Heterogeomys hispidus* is rather thin, long, and slightly curved. The pubis is also thin. The ischium is strongly developed. The acetabulum is deep. The pelvis of *Liomys salvini* has a moderatley expanded and curved ilium. The ischium and pubis are rather strongly structured. The acetabulum is proportionately small. The ilium of *Coendou mexicanus* is rather broad and slightly curved. The pubis is thin, having a curved apex. The ischial tuberosity is well developed. The acetabulum is relatively large and deep. The pelvis of *Agouti paca* is noticeably long and thin. The ilium is long, thin, and slightly expanded. The pubic spine and ischial process are well developed. The pubis is thin. The pelvis of *Dasyprocta punctata* is noticeably compressed in the area of the pubis and ischium, producing a long, oval obturator foramen. The ilium is nearly the same form as that of *Agouti paca*.

Order: Carnivora — Procyonids and Mustelids *(fig. 60)*

The pelvis of *Potos flavus* is short and heavy. The ilium is notched on both sides at its base. The ischium and pubis are about equal in size. Their common margin is rounded. The acetabulum and the obturator foramen are large. The pelves of both *Eira barbara* and *Galictis allamandi* are of about equal size and form. The ilia are heavy with little or no expansion or flattening. The ischia are also short and heavy. The pubes are moderately developed. The acetabula in both mustelids are quite large and deep. *Lutra annectens* has a pelvis that is heavier than any of the other mustelids. The ilium is expanded but short and heavy. The pubis and ischium are also heavily constructed. The acetabulum is unusually large and comparatively shallow.

Order: Perissodactyla — Tapirs *(fig. 61)*

The ilium of *Tapirus bairdii* is long and broad with an expanded crest. The pubis is considerably extended. The ischium is broadly flared. The acetabulum is round and deep. The acetabular fossa is widespread. The obturator foramen approaches a circular form.

Order: Artiodactyla — Peccaries and Deer *(fig. 61)*

Both peccaries, *Tayassu tajacu* and *Tayassu pecari*, have pelves that are quite similar. The main difference is the smaller size of *Tayassu tajacu*. The ilia are considerably curved with slightly expanded distal ends. The pubes are thinly constructed. The ischia are strongly built with large, oval obturator foramina and deep acetabula with pinched acetabular fossae. The two deer, as with their other bones, are separated by size, *Mazama americana* being considerably smaller than *Odocoileus virginianus*. In both, the ilia are long and thin with expanded terminal ends. The outlines of the combined ischia and pubes differ between the two cervids. There is a decided step in the margin of the ischium in *Mazama americana*. There is a gently curving margin in *Odocoileus virginianus*. At the union of the ascending ramus of the ischium with the descending ramus of the pubis, a right angle is formed in *Mazama americana*. This same area is a rounded margin in *Odocoileus virginianus*.

Femur

Order: Marsupialia — Opossums *(fig. 62)*

In the opossums, *Philander opossum*, *Marmosa mexicana*, and *Caluromys derbianus*, the femora have moderately strong, straight shafts. The lesser trochanters are well developed. The greater trochanters are substantial except that in *Marmosa mexicana* which is decidedly blunt. In this opossum the head lies close to the proximal end of the shaft without a noticeable neck. The lateral condyles in *Caluromys derbianus* and *Philander opossum* have decided outward extensions. Such an extension is not present in *Marmosa mexicana*.

Order: Chiroptera — Bats *(fig. 62)*

The small size and fragile construction of bat femora easily suggest these animals. The features that are peculiar to chiropteran femora are the nearly equal size of the trochanters and their arrangement with the femoral head. The distal condyles are weakly formed because of their lack of functional importance. The shaft of this element in *Desmodus rotundus* is quite compressed or flattened for the entire length of the shaft. It is thin, straight, and round in cross section in *Artibeus lituratus*.

Order: Primates — Monkeys *(figs. 62, 63)*

The femur of *Ateles geoffroyi* is considerably longer than that of *Alouatta villosa*. In both the shafts are long, thin, and nearly straight. The heads in both monkeys are quite round with noticeably restricted necks. The greater trochanters are well developed and project to about the same levels as the heads. The lesser tuberosities are stout, knobby processes.

Order: Edentata — Anteaters and Sloths *(fig. 63)*

The femur of *Myrmecophaga tridactyla*, as with the other anteaters, is heavy and wide but considerably flattened or compressed. There is a prominent, sharp ridge running the length of the femur from the greater trochanter to the lateral condyle. There is no developed lesser condyle. The head, as with most edentates, is huge in comparison with the rest of the femur and to proportions of this element in other mammals. In *Tamandua tetradactyla*, the shaft of the femur is more rounded in cross section, and there is no prominent crest on the lateral face. The head is large, rising considerably above the greater trochanter. The lesser trochanter is present as a small but prominent knob. The small size of *Cyclopes didactylus* sets it apart from the other anteaters. The femur, as in *Myrmecophaga tridactyla*, is considerably flattened in cross section. The head is huge and seems out of proportion with the rest of the bone. There is little or no neck. There is no defined ridge on the shaft, but the lateral margin is sharp and angular. Both the greater and lesser tuberosities are well developed. Of the two sloths, *Bradypus griseus* has the smaller femur. It is slightly compressed, resulting in an oval cross section. There is little or no neck. There is no lateral crest, but the lateral margin of the bone is sharp and angular. The head is large and rises above the greater trochanter, which is quite reduced in size. The lesser trochanter is well developed. The femoral shaft of *Choloepus hoffmanni* is long and thin and nearly round in cross section. There is a slight ridge or crest on the proximal end of the shaft on the lateral face. The head is large with a well-defined neck. It rises well above the greater trochanter which is quite undeveloped. The lesser trochanter is well defined and prominent.

Order: Rodentia — Rodents *(fig. 64)*

The femur of *Heterogeomys hispidus* has a noticeable enlargement of all its processes and condyles. The head, greater and lesser trochanters, and distal articular condyles are all quite large. There is a noticeable flaring of the shaft above and joining with the medial and lateral condyles. The proximal end of the bone is flattened, producing a crest on the lateral face of the femur opposite the lesser trochanter. In *Liomys salvini,* this lateral crest is reduced. The head appears to be too small in relation to the rest of the femur. The greater trochanter is rather well represented, but the lesser trochanter is less apparent. There is no widening of the shaft in the area of the medial and lateral condyles. The femur of *Coendou mexicanus* is noticeably flattened in cross section. The shaft is strong and heavy. The head and greater and lesser trochanters are well developed. The head is set off by a defined neck. In *Agouti paca,* the lateral margin of the femoral shaft, which is heavy and strong, has a moderately developed crest. The huge greater tuberosity rises well above the head, which also has a defined neck separating it from the shaft. The lesser tuberosity is slight. The distal articular condyles are well developed. In *Dasyprocta punctata,* the shaft of the femur is well developed but lighter in build than in the other two large rodents just described. The lateral face of the shaft has a sharp ridge running its length. The heavy greater trochanter rises above the head, which like the other rodents, is set off from the shaft by a well-defined neck. The lesser trochanter is a mere knob of bone.

Order: Carnivora — Procyonids and Mustelids *(figs. 64, 65)*

The femur of *Potos flavus* has a head and neck that are noticeably large and heavy. The shaft is long and thin and round in cross section. The greater trochanter is moderately developed. The lesser trochanter is present as a rather small process. In *Eira barbara,* the femoral shaft is long and thin and rounded in section. The head is quite strong with no prominent neck. The greater trochanter is substantial and is extended to the same level as the head. The distal articular condyles are strongly developed. In *Galictis allamandi,* the femoral shaft is more heavily constituted than in *Eira barbara.* It has a decided curve about midway on the lateral face of the shaft. The greater trochanter is quite low. The head is enlarged and rises well above the greater trochanter. There is no defined neck. The lesser trochanter is little developed. The distal condyles are quite large and strongly constructed. In *Lutra annectens,* the entire bone is strong, wide, and heavily constructed. The shaft, although compressed somewhat, is short and heavy. The head and greater trochanter are strong. The lesser tuberosity is present as a roughened knobby process. The distal articular condyles are enlarged to be almost out of proportion with the rest of the femur.

Order: Perissodactyla — Tapirs *(fig. 65)*

The femur of *Tapirus bairdii* can be distinguished by a single process if that portion of the bone is present. This is the third trochanter that arises from the lateral face of the shaft between the greater trochanter and the lateral articular condyle. It is characteristic of perissodactyls. The greater trochanter and lesser trochanter are quite prominent and well represented. The latter is present as a short, high crest. The head is set off by a stout neck. The distal articular condyles are strongly developed. This femur will be the largest of those found in the pre-Columbian Maya faunal assemblages.

Order: Artiodactyla — Peccaries and Deer *(figs. 65, 66)*

As with some of the other postcranial elements, the femora of both *Tayassu tajacu* and *Tayassu pecari* are not readily separable as to which species they represent. They will be treated as one. The shaft is straight and rather strong. The greater trochanter is strong but low. The head is rather short and set close to the shaft. The lesser trochanter is an enlargement at the terminus of a ridge running from the greater tuberosity to the medial face of the shaft. It is not prominent. There is a slightly developed medial epicondyle. The deer can be separated by size alone, the brocket being considerably smaller than the white-tailed deer. The shafts of both deer are strongly constructed and round in section. In each species, the proximal end of the bone is most diagnostic. The greater tuberosity is connected to the head by a ridge that blends into the articular face with no shoulder, lip, or break in the margin. The undersurface of the articular surface of the head also joins the shaft in a gentle curve rather than in a defined lip as in most other mammal femora. The trochanteric ridge is prominent, and there is a slightly developed lesser trochanter at its base. The distal articular condyles and trochlea are well developed.

Tibia and Fibula

Order: Marsupialia — Opossums *(fig. 66)*

The tibiae and fibulae are separate bones in the opossum. The tibiae are somewhat flattened and curved in *Philander opossum, Marmosa mexicana,* and *Caluromys derbianus.* Both articular ends in all three opossums are well developed. The form of the fibulae in these small marsupials is striking. The shafts are thin with enormous expansion and flattening of the proximal ends. There are three distinct processes on the distal margins of these widened extremities.

Order: Chiroptera — Bats *(fig. 67)*

The tibiae and fibulae of bats in general are weakly constructed and functionally not of first importance to the locomotion of these small mammals. The fibulae of both *Artibeus lituratus* and *Desmodus rotundus* lie in close contact with the tibiae along the entire length, held in close contact with soft connecting tissues. Both are considerably flattened in the vampire.

Order: Primates — Monkeys *(fig. 67)*

The tibia and fibula of *Ateles geoffroyi* are considerably larger than these same bones in *Alouatta villosa.* The tibiae have well-developed heads in both monkeys and extended medial malleoli in both. The tibial shafts are also a bit curved. The fibulae in both monkeys have slender shafts, widened heads, and well-developed lateral malleoli.

Order: Edentata — Anteaters and Sloths *(figs. 67, 68)*

The tibia of *Myrmecophaga tridactylus* is stout and heavy with a heavily ridged shaft. The proximal and distal ends of the tibia are quite expanded. The medial malleolus is strongly developed. The fibular shaft is flattened and slightly twisted. Both proximal and distal ends are flattened. The tibia of *Tamandua tetradactyla* has a strong shaft with a round cross section. The proximal end of the shaft is slightly curved, causing the lateral end to overhang slightly. The fibula is slender with no distinguishing processes. The smallest anteater, *Cyclopes didactylus,* can be separated by size alone. The tibia and fibula are somewhat compressed. The fibula is considerably flattened at its proximal end. The distal ends of both elements are closely united and at times even fused. The lateral malleolus is slightly turned outward from the line of the shaft. The smaller size of *Bradypus griseus* distinguishes its tibia and fibula from that of *Choloepus hoffmanni.* The two bones of *Bradypus griseus* are heavy with strongly developed processes. The tibial shaft is considerably flattened with a moderate medial malleolus. The fibula is considerably curved, having a ball-shaped lateral malleolus. The shaft of the tibia of *Choloepus hoffmanni* is thin and curved with strong articular processes. The fibula is quite thin at the prox-imal end. The distal end is expanded, having a hooked lateral malleolus.

Order: Rodentia — Rodents *(figs. 68, 69)*

Rodents show considerable tibial and fibular variation. Some genera have separate tibiae and fibulae. Others have fused tibiae and fibulae. In both *Heterogeomys hispidus* and *Liomys salvini,* the tibiae and fibulae are joined to common shafts from the midpoints to the distal ends. In each the fused proximal ends of both the tibia and fibula result in an arched feature of this combination element. In *Coendou mexicanus,* the head is considerably expanded with a crest running from between the condyles to a point half way down the shaft. The fibula is strong and curved, with a flattened proximal end and a strong distal articular condyle. The proximal ends of both bones are often lightly fused together. The tibia of *Agouti paca* has a heavy, strong shaft which is considerably ridged. Both the proximal and distal articular condyles are well developed. The fibula is thin and slightly twisted. The proximal end is greatly flattened. The lateral malleolus is a strongly developed process. The tibia of *Dasyprocta punctata* is moderately heavy, smooth, and nearly straight. The proximal condyle is well developed, as is the medial malleolus. The fibula is a thin splint of bone with a flattened, spatulate proximal end. The lateral malleolus is well defined for so slender an element.

Order: Carnivora — Procyonids and Mustelids *(fig. 69)*

The tibia of *Potos flavus* has a widely expanded head. The shaft is round in cross section. The medial malleolus is strongly developed. The fibula is thin and straight with well-developed proximal and lateral condyles. The tibia of *Eira barbara* has a strong shaft. The medial malleolus is a slightly rounded process not greatly extended. The fibula is thin, round, and straight. Both proximal and distal condyles are well developed. In *Galictis allamandi,* the tibial shaft is slightly curved with a strong, wide articular head and a medial malleolus that is strong with a downward extension. The fibula is moderately heavy with a slight ridge on the proximal end of the shaft. The lateral malleolus is quite heavily developed. In *Lutra annectens,* the tibia is quite heavy with an unusually strong head and a heavy, extended medial malleolus. The fibula is straight and strongly constructed. The lateral malleolus is enormous and club-shaped.

Order: Perissodactyla — Tapirs *(fig. 70)*

The tibia and fibula are separate and well developed in *Tapirus bairdii.* The shaft of the tibia is moderately heavy with strong proximal and distal articular condyles. The medial malleolus is well developed. The fibula is slightly flattened and straight with moderately developed articular ends.

Order: Artiodactyla — Peccaries and Deer *(fig. 70)*

The tibiae and fibulae of both *Tayassu tajacu* and *Tayassu pecari* may be separate or partially fused at their ends. The shafts of the tibiae are heavy, slightly curved, and moderately ridged. The processes on both the proximal and distal ends are well developed. The fibulae are separate elements and are long and thin with triangular cross sections. The proximal and distal articulations are expanded and somewhat flattened. In both *Odocoileus virginianus* and *Mazama americana*, the shafts are nearly straight and lightly ridged. The medial malleoli are moderately extended. The fibulae are present as slight vestigial spurs extending down from the outer margin of the lateral condyle.

Order: Sirenia — Manatees *(fig. 71)*

In general the skeleton of *Trichechus manatus* is structurally different from any other mammal. There is little or no cancellous or marrow tissue in the limb bones or ribs. Instead they have a solid, ivorylike composition that is termed pachyostosis. These animals are generally identified from rib fragments alone because of the numerous diagnostic fragments preserved from destruction by the elements and also because of the selection of these animals as a tool and artifact source by various cultures within their range.

Scapula

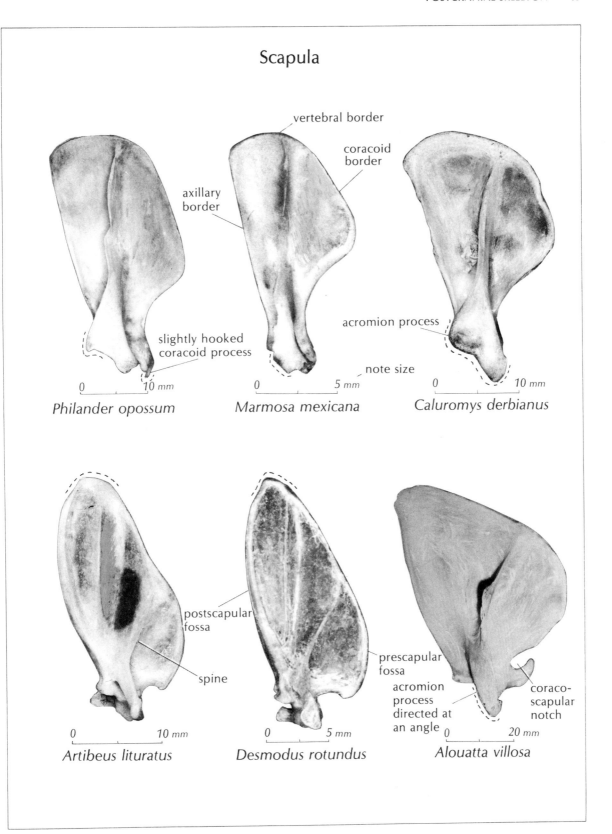

FIGURE 44

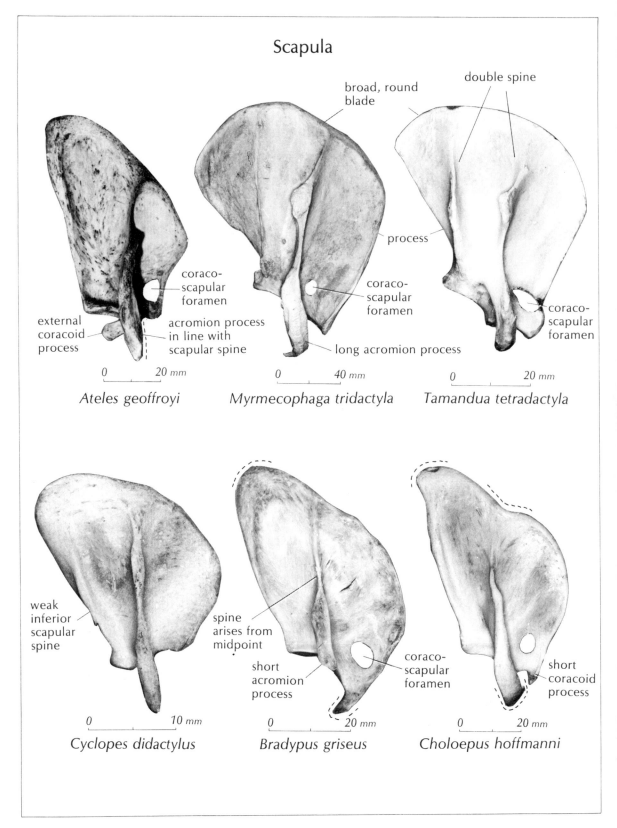

Scapula

Ateles geoffroyi

Myrmecophaga tridactyla

Tamandua tetradactyla

Cyclopes didactylus

Bradypus griseus

Choloepus hoffmanni

FIGURE 45

Scapula

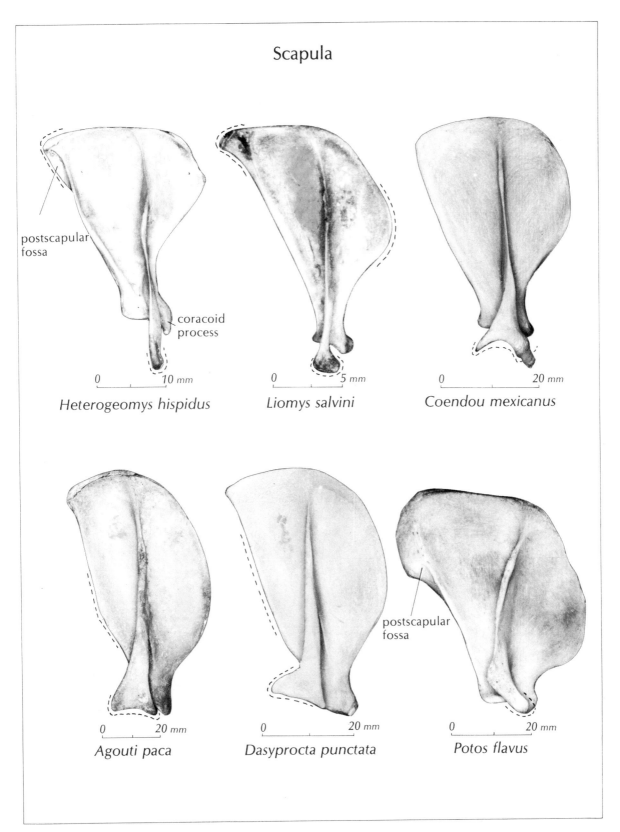

postscapular
fossa

coracoid
process

0   10 mm
Heterogeomys hispidus

0   5 mm
Liomys salvini

0   20 mm
Coendou mexicanus

0   20 mm
Agouti paca

0   20 mm
Dasyprocta punctata

postscapular
fossa

0   20 mm
Potos flavus

FIGURE 46

Scapula

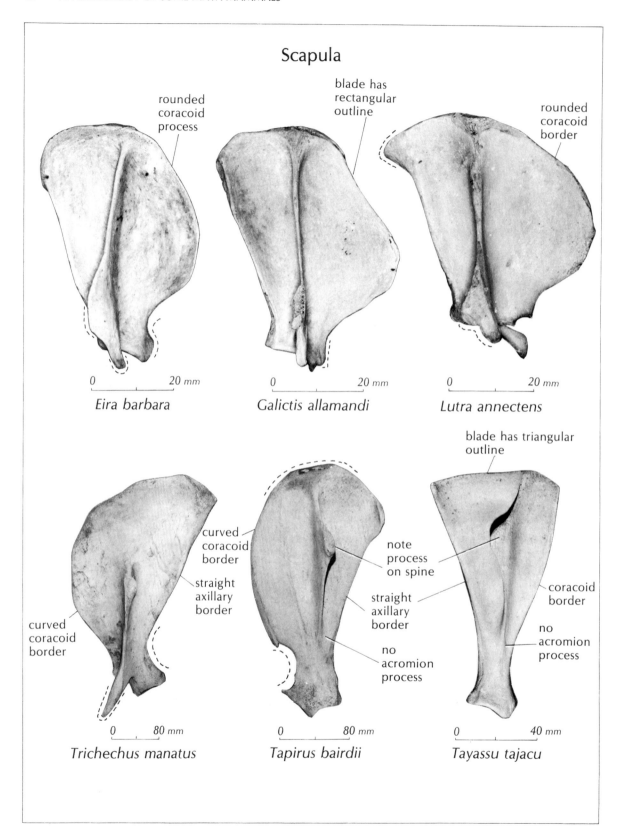

rounded coracoid process

blade has rectangular outline

rounded coracoid border

Eira barbara

Galictis allamandi

Lutra annectens

0   20 mm

0   20 mm

0   20 mm

blade has triangular outline

curved coracoid border

straight axillary border

curved coracoid border

note process on spine

straight axillary border

coracoid border

no acromion process

no acromion process

Trichechus manatus

Tapirus bairdii

Tayassu tajacu

0   80 mm

0   80 mm

0   40 mm

FIGURE 47

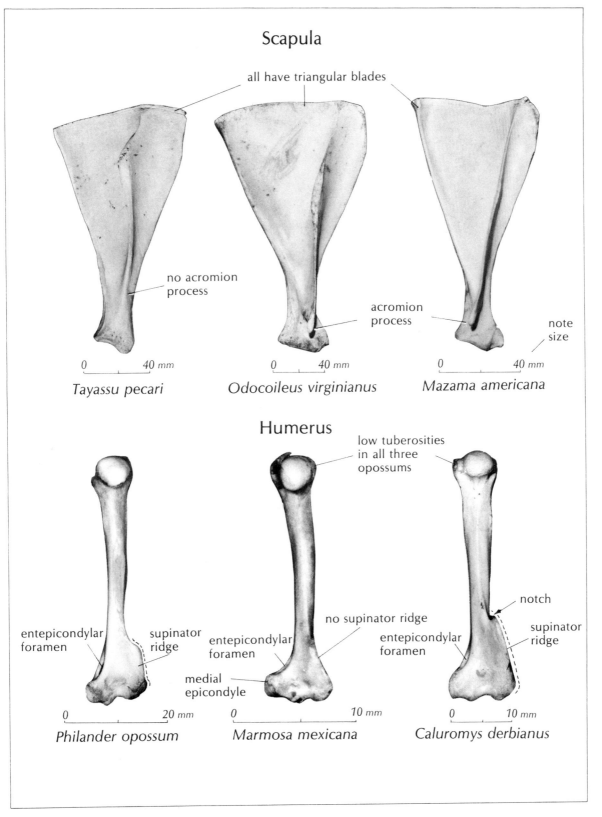

Scapula

all have triangular blades

no acromion process

acromion process

note size

0   40 mm   0   40 mm   0   40 mm

Tayassu pecari   *Odocoileus virginianus*   *Mazama americana*

Humerus

low tuberosities in all three opossums

entepicondylar foramen

supinator ridge

no supinator ridge

entepicondylar foramen

medial epicondyle

entepicondylar foramen

notch

supinator ridge

0   20 mm   0   10 mm   0   10 mm

Philander opossum   *Marmosa mexicana*   *Caluromys derbianus*

FIGURE 48

Humerus

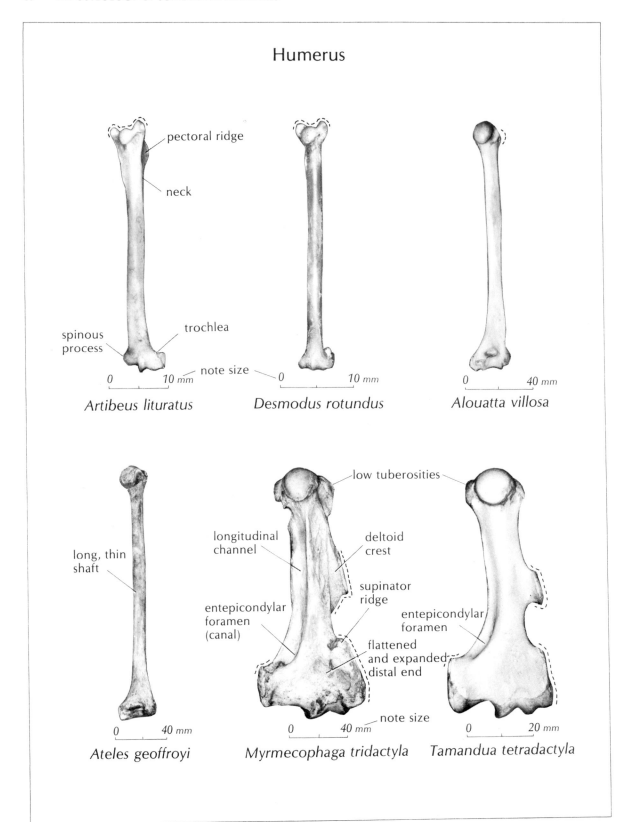

pectoral ridge

neck

spinous process

trochlea

note size

0   10 mm

Artibeus lituratus

0   10 mm

Desmodus rotundus

0   40 mm

Alouatta villosa

long, thin shaft

longitudinal channel

entepicondylar foramen (canal)

low tuberosities

deltoid crest

supinator ridge

flattened and expanded distal end

entepicondylar foramen

note size

0   40 mm

Ateles geoffroyi

0   40 mm

Myrmecophaga tridactyla

0   20 mm

Tamandua tetradactyla

FIGURE 49

Humerus

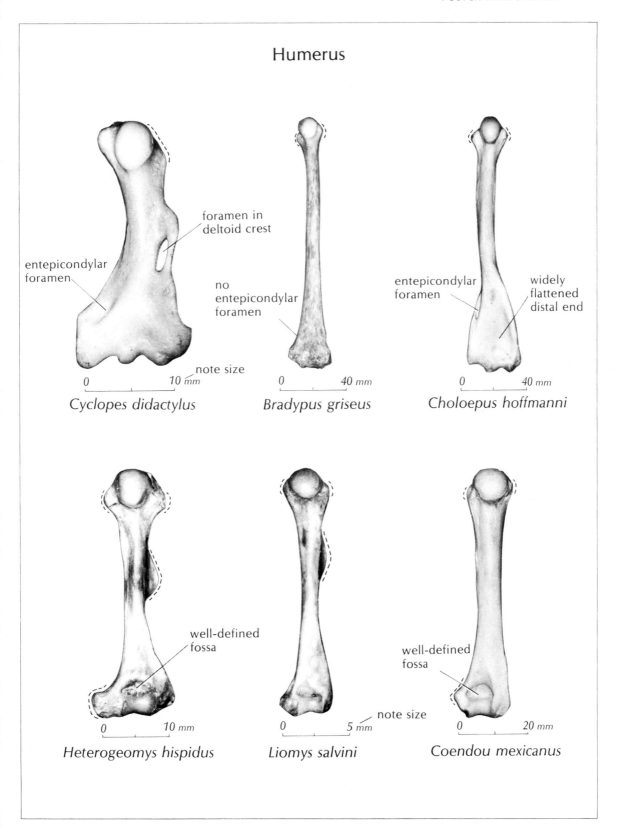

Cyclopes didactylus

entepicondylar foramen

foramen in deltoid crest

note size

0   10 mm

Bradypus griseus

no entepicondylar foramen

0   40 mm

Choloepus hoffmanni

entepicondylar foramen

widely flattened distal end

0   40 mm

Heterogeomys hispidus

well-defined fossa

0   10 mm

Liomys salvini

note size

0   5 mm

Coendou mexicanus

well-defined fossa

0   20 mm

FIGURE 50

Humerus

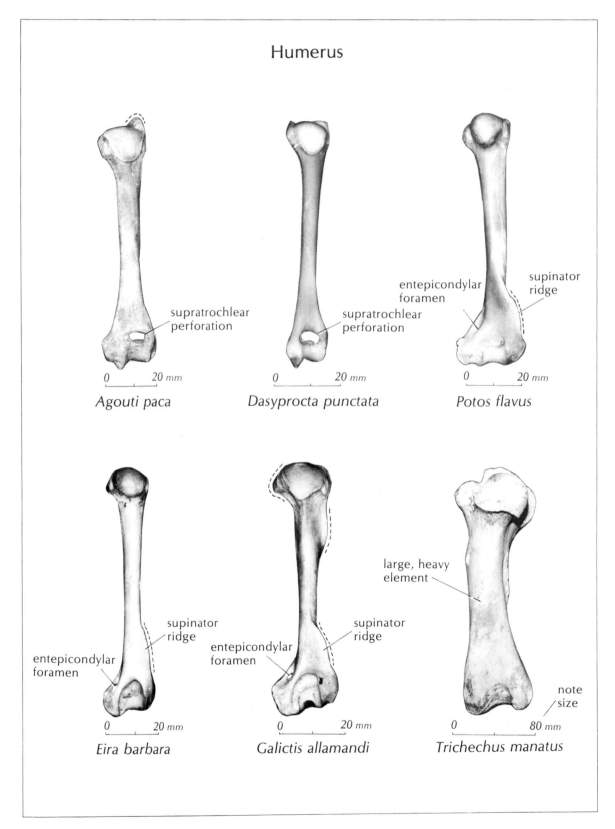

supratrochlear perforation

Agouti paca

supratrochlear perforation

Dasyprocta punctata

entepicondylar foramen

supinator ridge

Potos flavus

entepicondylar foramen

supinator ridge

Eira barbara

entepicondylar foramen

supinator ridge

Galictis allamandi

large, heavy element

note size

Trichechus manatus

FIGURE 51

Humerus

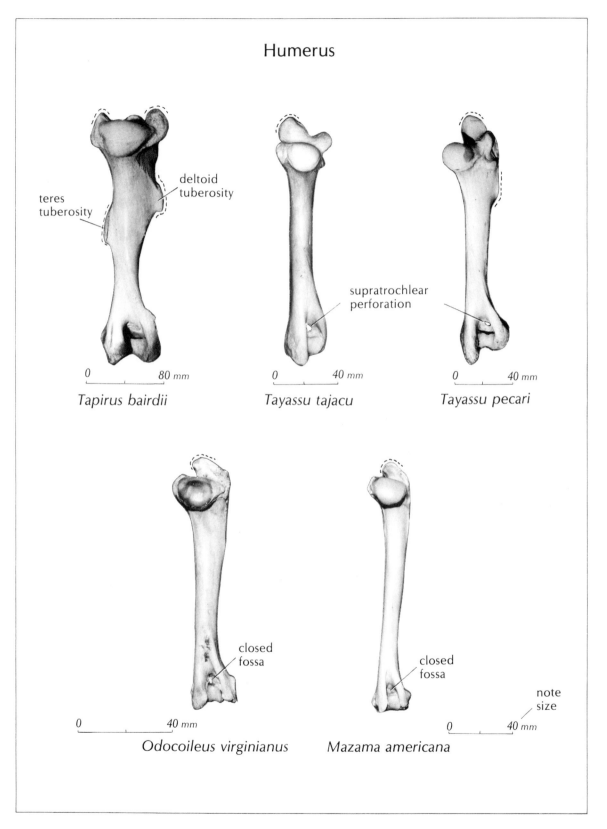

teres tuberosity

deltoid tuberosity

supratrochlear perforation

Tapirus bairdii

0   80 *mm*

Tayassu tajacu

0   40 *mm*

Tayassu pecari

0   40 *mm*

closed fossa

closed fossa

0   40 *mm*

Odocoileus virginianus

0   40 *mm*

note size

Mazama americana

FIGURE 52

Radius and Ulna

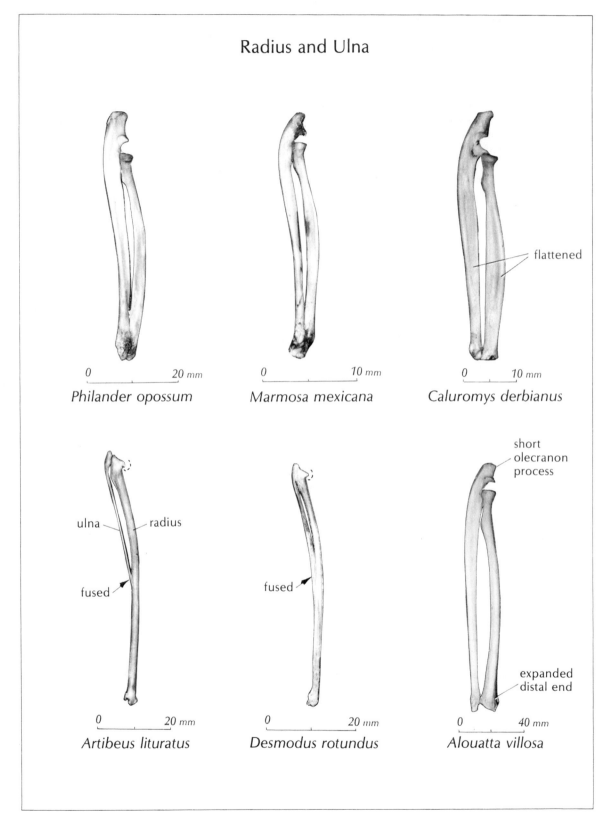

0 —————— 20 mm
Philander opossum

0 —————— 10 mm
Marmosa mexicana

flattened

0 —————— 10 mm
Caluromys derbianus

ulna radius

fused

0 —————— 20 mm
Artibeus lituratus

fused

0 —————— 20 mm
Desmodus rotundus

short olecranon process

expanded distal end

0 —————— 40 mm
Alouatta villosa

FIGURE 53

Radius and Ulna

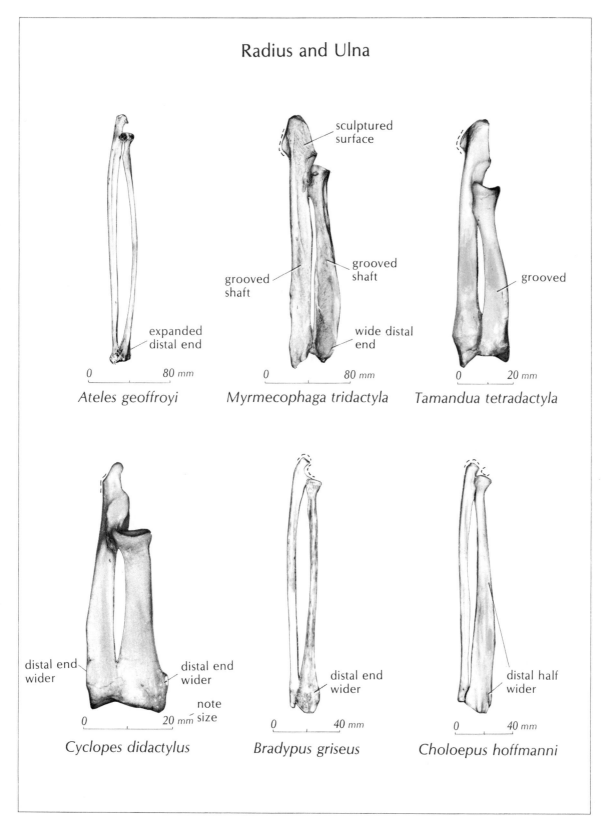

FIGURE 54

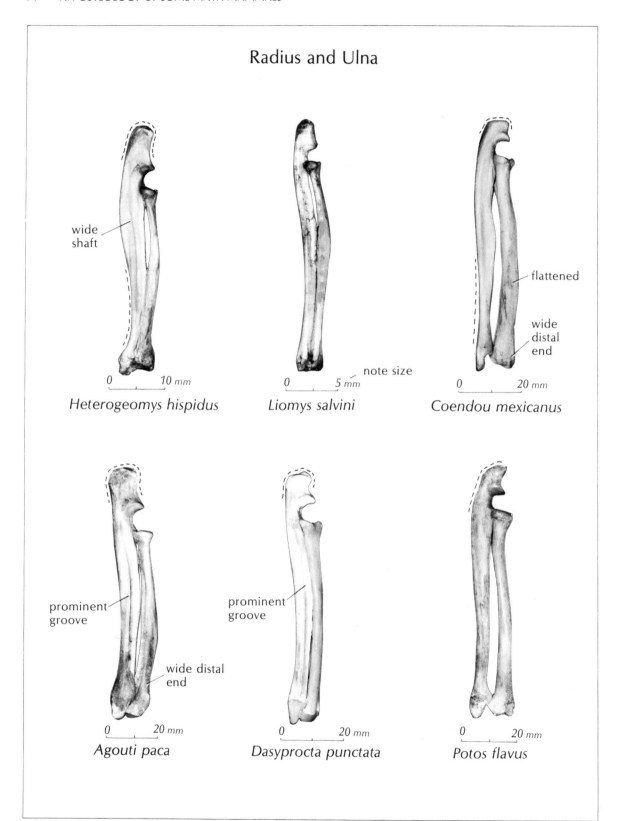

Radius and Ulna

wide shaft

Heterogeomys hispidus

note size

Liomys salvini

flattened

wide distal end

Coendou mexicanus

prominent groove

wide distal end

Agouti paca

prominent groove

Dasyprocta punctata

Potos flavus

FIGURE 55

Radius and Ulna

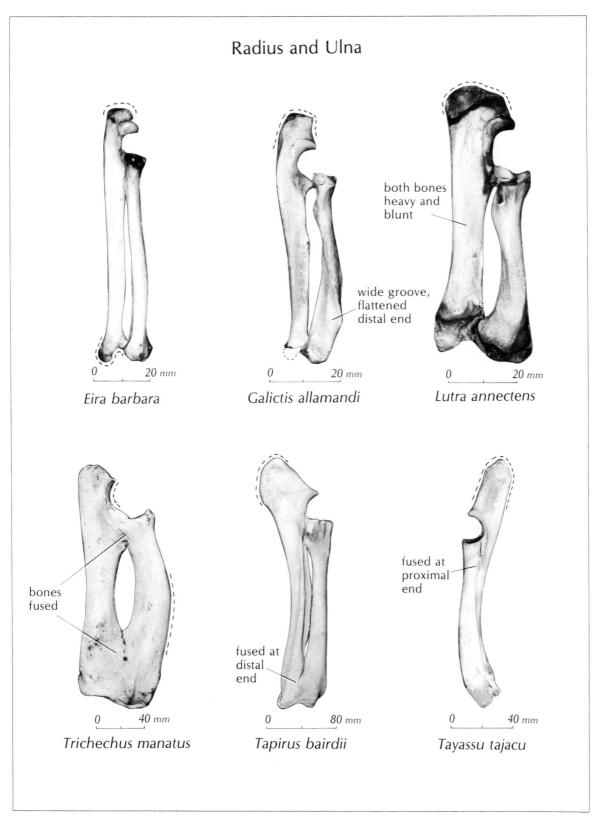

Eira barbara

Galictis allamandi

both bones
heavy and
blunt

wide groove,
flattened
distal end

Lutra annectens

bones
fused

Trichechus manatus

fused at
distal
end

Tapirus bairdii

fused at
proximal
end

Tayassu tajacu

FIGURE 56

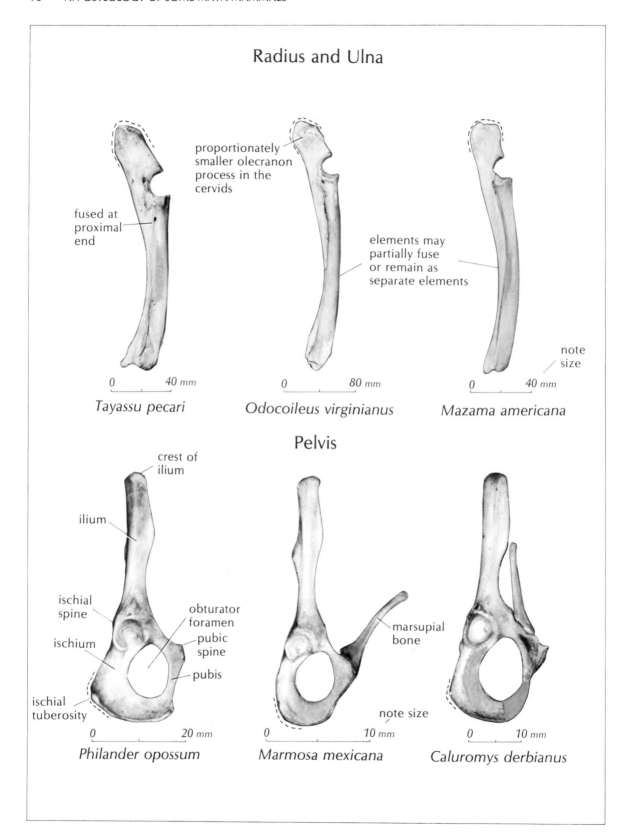

Radius and Ulna

fused at proximal end

proportionately smaller olecranon process in the cervids

elements may partially fuse or remain as separate elements

note size

0   40 mm

0   80 mm

0   40 mm

Tayassu pecari

Odocoileus virginianus

Mazama americana

Pelvis

crest of ilium

ilium

ischial spine

ischium

obturator foramen

pubic spine

pubis

ischial tuberosity

marsupial bone

note size

0   20 mm

0   10 mm

0   10 mm

Philander opossum

Marmosa mexicana

Caluromys derbianus

FIGURE 57

Pelvis

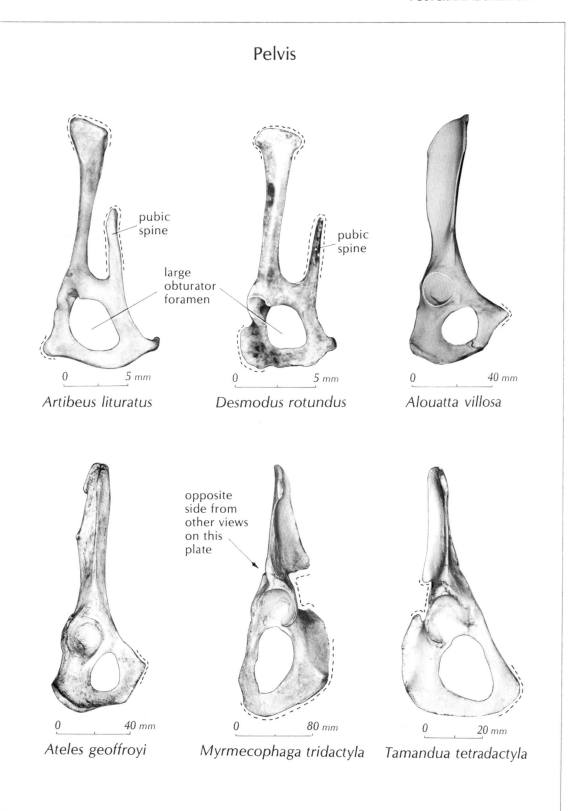

pubic spine

large obturator foramen

pubic spine

Artibeus lituratus

0   5 mm

Desmodus rotundus

0   5 mm

Alouatta villosa

0   40 mm

opposite side from other views on this plate

Ateles geoffroyi

0   40 mm

Myrmecophaga tridactyla

0   80 mm

Tamandua tetradactyla

0   20 mm

FIGURE 58

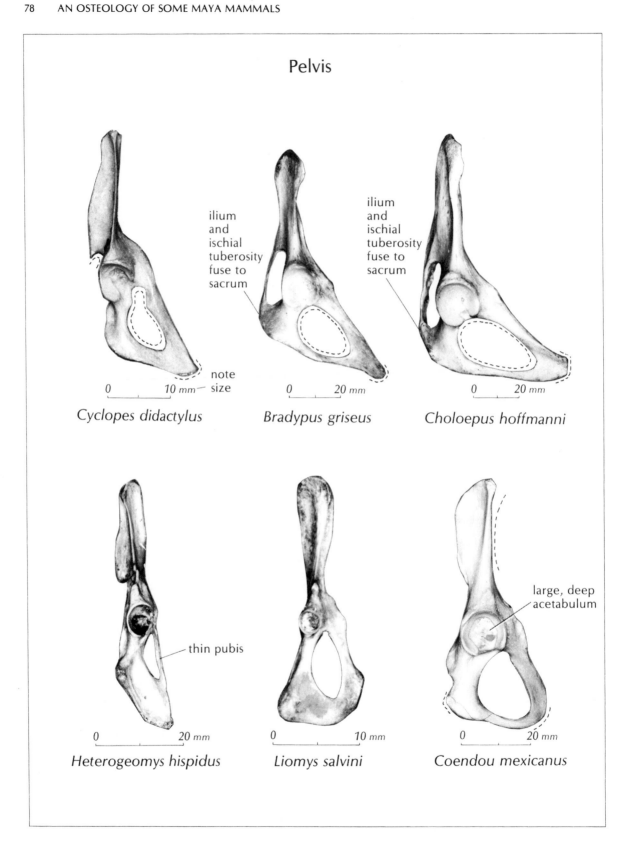

Pelvis

ilium and ischial tuberosity fuse to sacrum

ilium and ischial tuberosity fuse to sacrum

note size

Cyclopes didactylus

Bradypus griseus

Choloepus hoffmanni

thin pubis

large, deep acetabulum

Heterogeomys hispidus

Liomys salvini

Coendou mexicanus

FIGURE 59

Pelvis

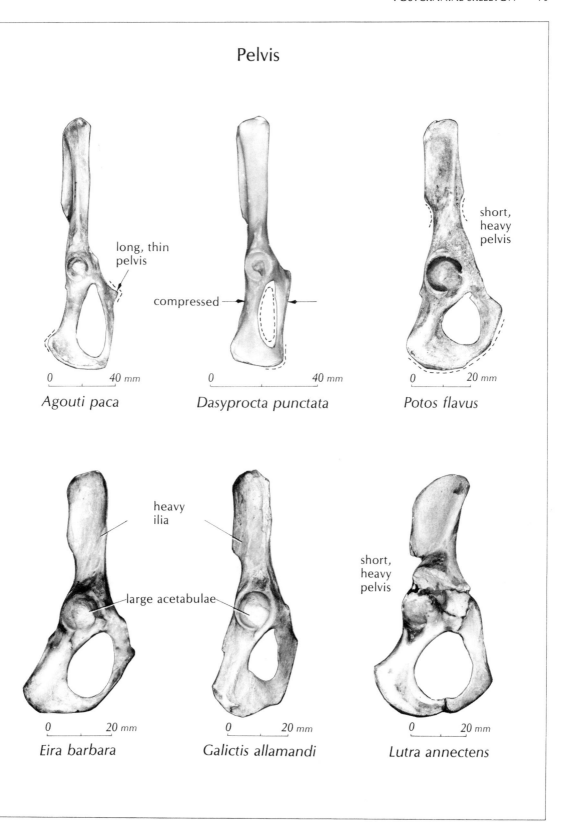

long, thin
pelvis

compressed

short,
heavy
pelvis

0   40 mm

0   40 mm

0   20 mm

Agouti paca

Dasyprocta punctata

Potos flavus

heavy
ilia

large acetabulae

short,
heavy
pelvis

0   20 mm

0   20 mm

0   20 mm

Eira barbara

Galictis allamandi

Lutra annectens

FIGURE 60

Pelvis

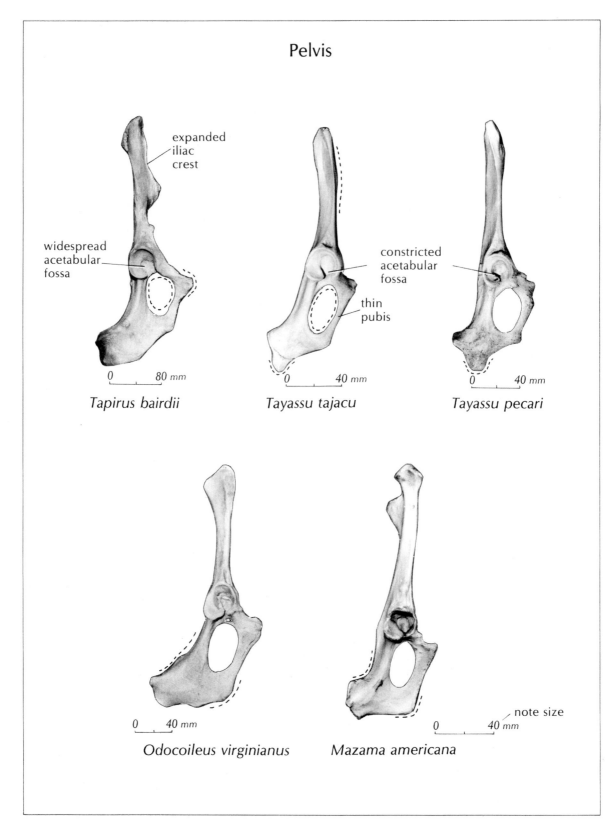

FIGURE 61

Femur

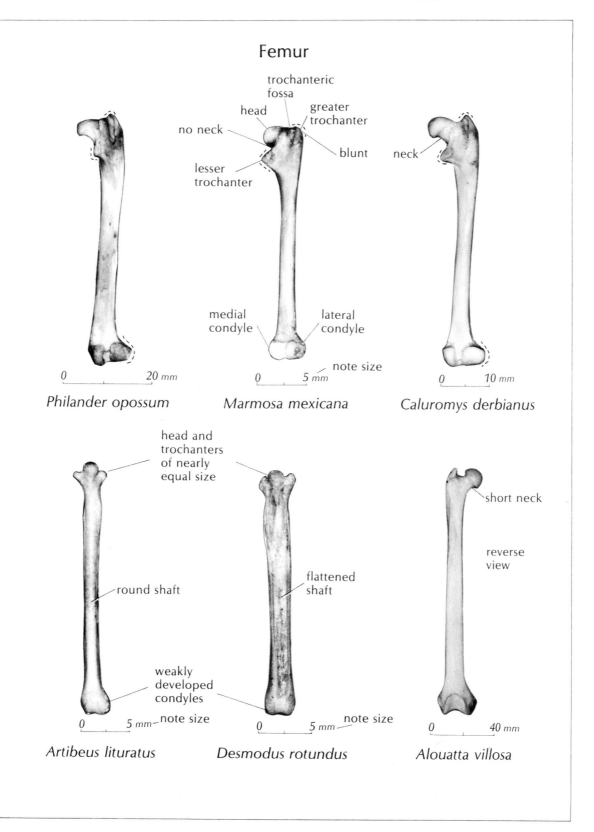

trochanteric
fossa

head

greater
trochanter

no neck

blunt

lesser
trochanter

neck

medial
condyle

lateral
condyle

note size

0   20 mm

Philander opossum

0   5 mm

Marmosa mexicana

0   10 mm

Caluromys derbianus

head and
trochanters
of nearly
equal size

short neck

reverse
view

round shaft

flattened
shaft

weakly
developed
condyles

0   5 mm note size

Artibeus lituratus

0   5 mm note size

Desmodus rotundus

0   40 mm

Alouatta villosa

FIGURE 62

Femur

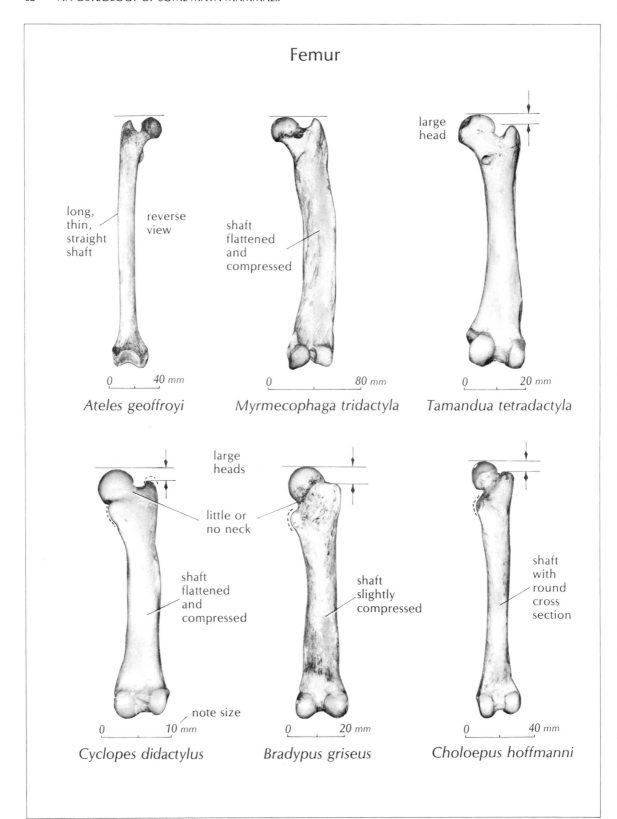

FIGURE 63

Femur

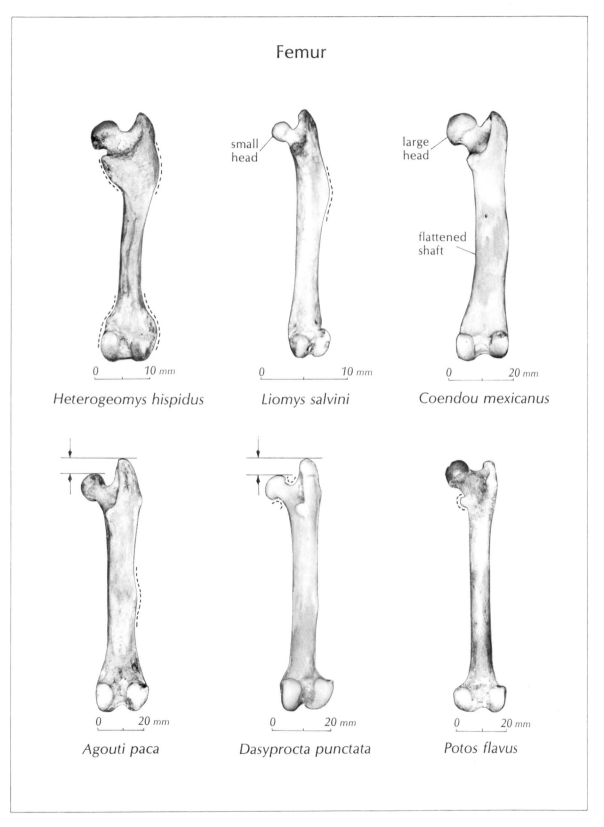

FIGURE 64

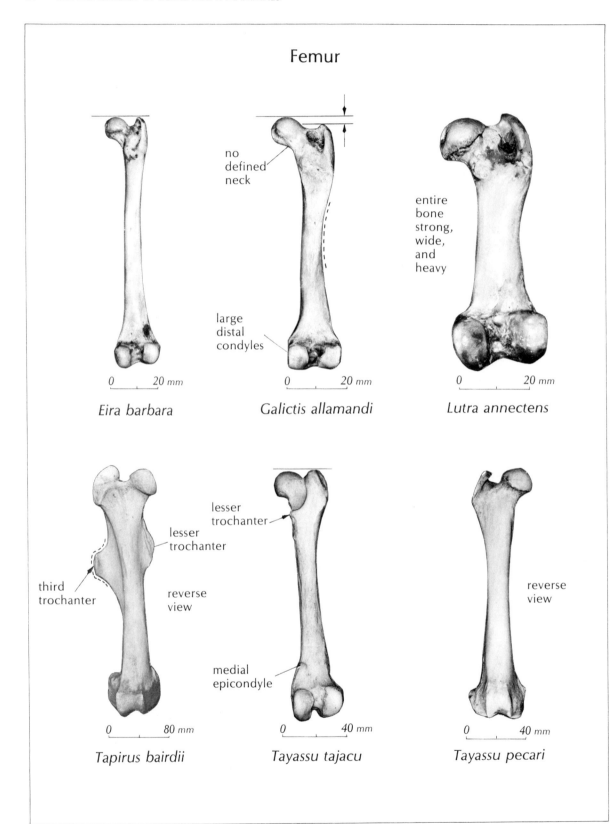

Femur

no defined neck

large distal condyles

entire bone strong, wide, and heavy

Eira barbara

Galictis allamandi

Lutra annectens

third trochanter

lesser trochanter

lesser trochanter

reverse view

medial epicondyle

reverse view

Tapirus bairdii

Tayassu tajacu

Tayassu pecari

FIGURE 65

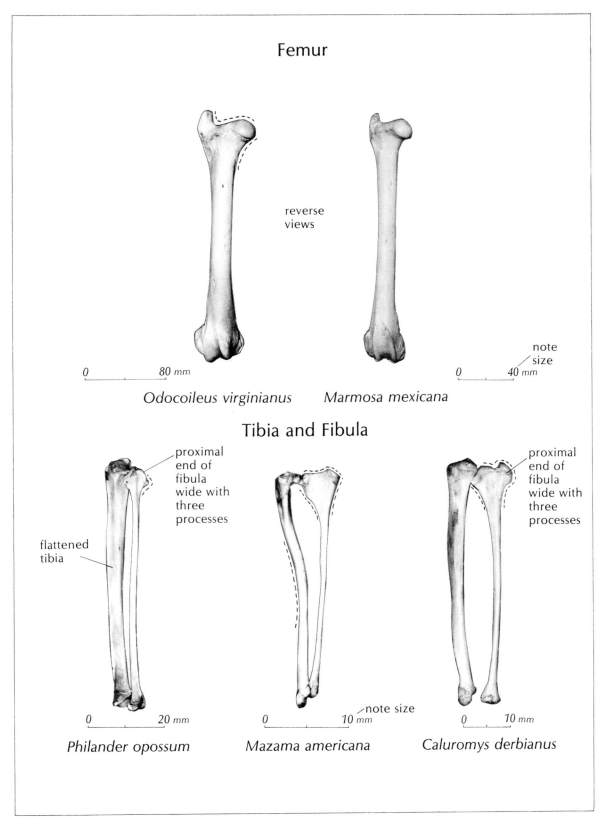

Femur

reverse
views

note
size

0   80 mm   0   40 mm

Odocoileus virginianus   *Marmosa mexicana*

Tibia and Fibula

proximal
end of
fibula
wide with
three
processes

flattened
tibia

proximal
end of
fibula
wide with
three
processes

0   20 mm   0   10 mm   0   10 mm
note size

Philander opossum   *Mazama americana*   *Caluromys derbianus*

FIGURE 66

Tibia and Fibula

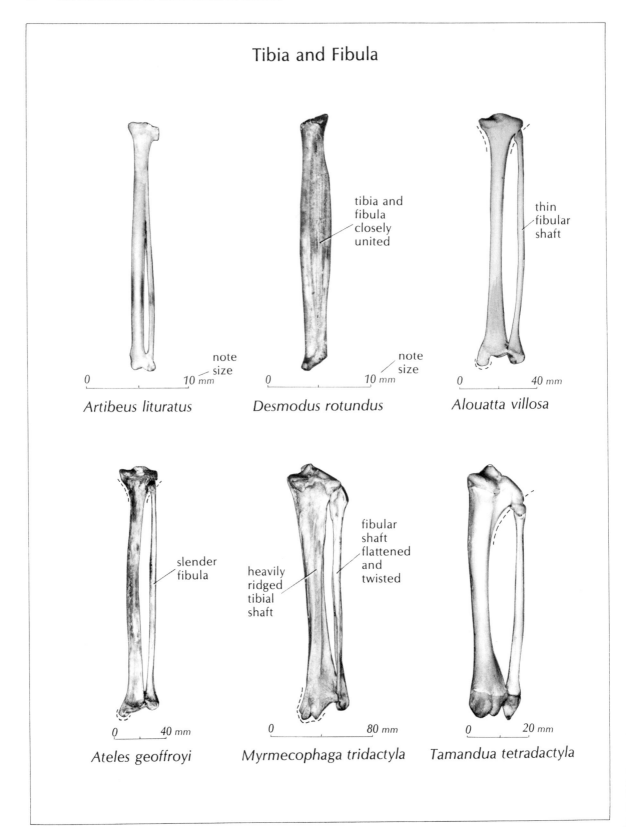

tibia and fibula closely united

note size

thin fibular shaft

Artibeus lituratus

Desmodus rotundus

Alouatta villosa

slender fibula

heavily ridged tibial shaft

fibular shaft flattened and twisted

Ateles geoffroyi

Myrmecophaga tridactyla

Tamandua tetradactyla

FIGURE 67

Tibia and Fibula

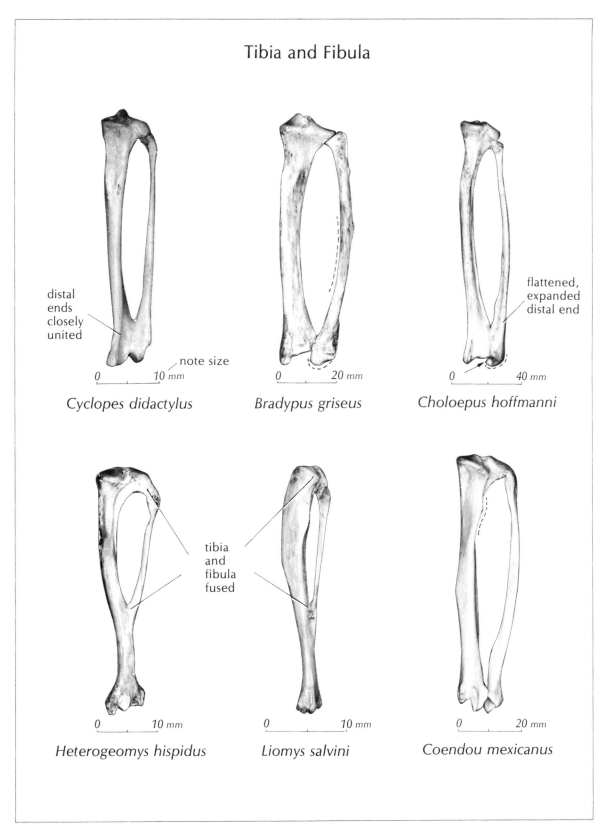

Cyclopes didactylus

Bradypus griseus

Choloepus hoffmanni

Heterogeomys hispidus

Liomys salvini

Coendou mexicanus

FIGURE 68

Tibia and Fibula

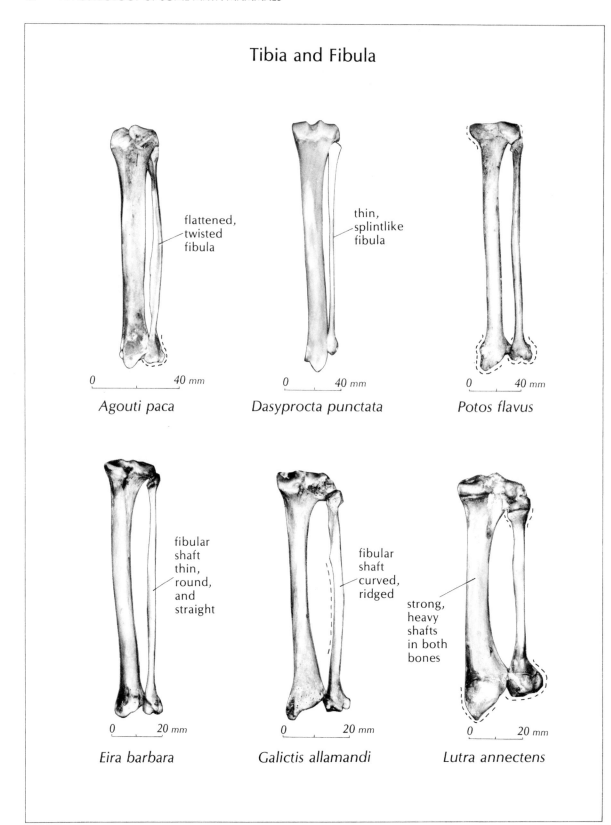

flattened, twisted fibula

thin, splintlike fibula

Agouti paca   *Dasyprocta punctata*   *Potos flavus*

fibular shaft thin, round, and straight

fibular shaft curved, ridged

strong, heavy shafts in both bones

Eira barbara   *Galictis allamandi*   *Lutra annectens*

FIGURE 69

Tibia and Fibula

separate tibiae and fibulae

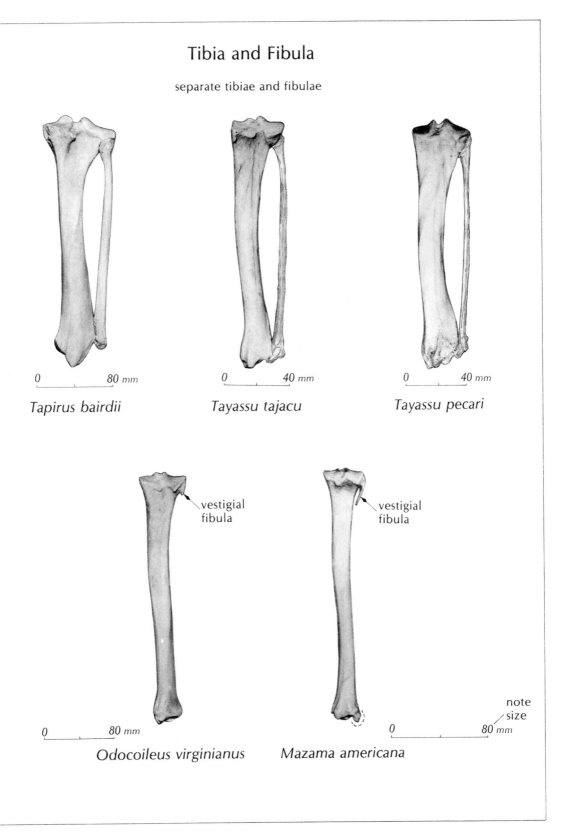

Tapirus bairdii

Tayassu tajacu

Tayassu pecari

Odocoileus virginianus

Mazama americana

FIGURE 70

Ribs and Cervical Vertebra of
Trichechus manatus

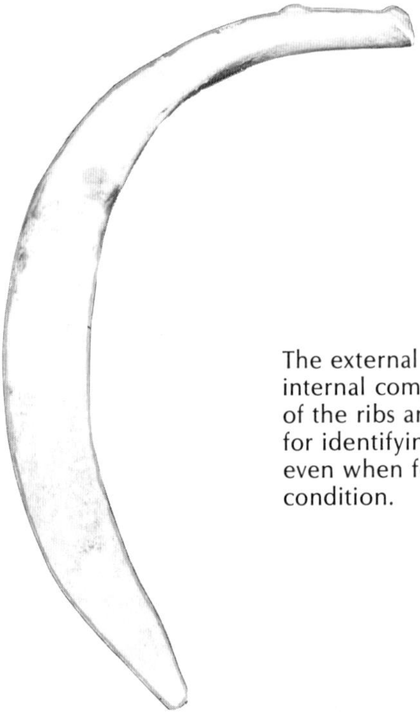

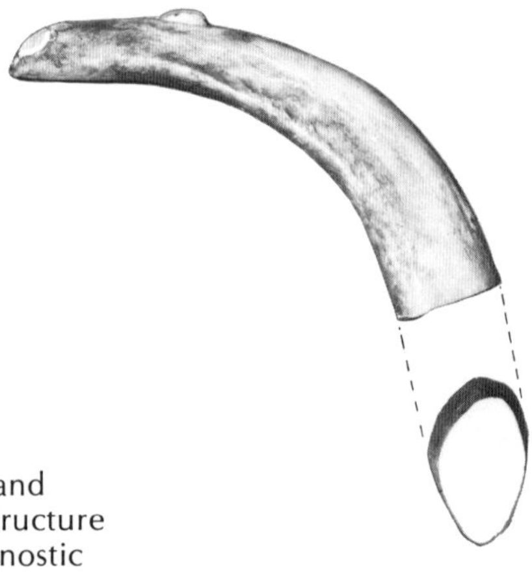

The external form and
internal compact structure
of the ribs are diagnostic
for identifying the manatee
even when found in fragmented
condition.

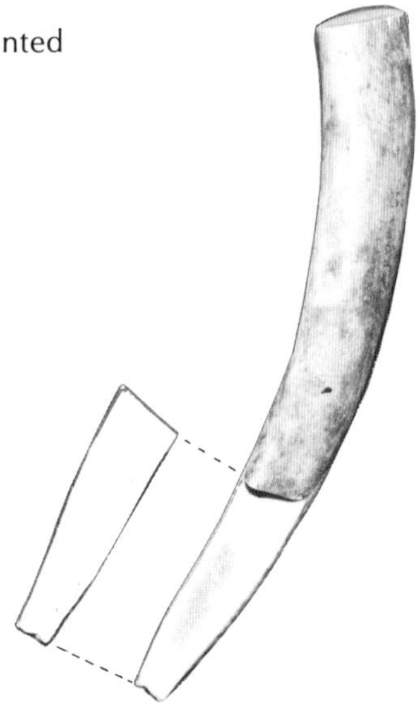

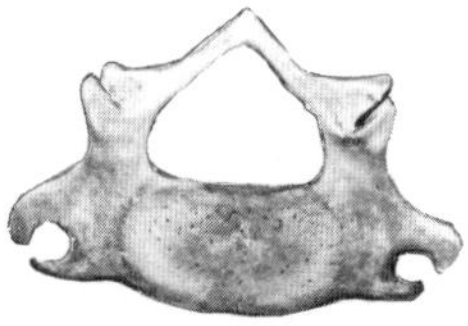

The cervical vertebra,
as in other aquatic
mammals, is consider-
ably more compressed
than that of terrestrial
animals of the same size.

0 —————————— 80 *mm*

FIGURE 71

References

Hall, E.R., and K.R. Kelson
  1959 *The Mammals of North America.* 2 vols. The Ronald Press Company, New York.

Leopold, A.S.
  1959 *Wildlife of Mexico.* University of California Press, Berkeley.

Olsen, S.J.
  1964 *Mammal Remains from Archaeological Sites: Part I, Southeastern and Southwestern United States.* Papers of the Peabody Museum, Harvard University, vol. 56, no. 1. Cambridge, Massachusetts.

  1972 "Animal Remains from Altar de Sacrificios," in Willey, G.R., *The Artifacts of Altar de Sacrificios.* Papers of the Peabody Museum, Harvard University, vol. 64, no. 1, pp. 243–246. Cambridge, Massachusetts.

  1978 "Vertebrate Faunal Remains," in Willey, G.R., *Excavations at Seibal: Artifacts.* Memoirs of the Peabody Museum, Harvard University, vol. 14, no. 1, pp. 172–176. Cambridge, Massachusetts.

Pollock, H.E.D., and C.E. Ray
  1957 "Notes on Vertebrate Animal Remains from Mayapan," *Current Reports,* no. 41, pp. 633–656. Carnegie Institute of Washington, Department of Archaeology, Washington, D.C.

Walker, E.P.
  1975 *Mammals of the World,* 3rd ed. 2 vols. Johns Hopkins University Press, Baltimore.

Woodburne, M.O.
  1968 *The Cranial Myology and Osteology of Dicotyles tajacu, the Collared Peccary, and its Bearing on Classification.* Memoirs of the Southern California Academy of Sciences, vol. 7. San Francisco.

Woodbury, R.B., and A.S. Trik
  1953 *The Ruins of Zaculeu, Guatemala.* vol. 1, pp. 277–278. A United Fruit Company project report. The William Byrd Press, Inc., Richmond, Virginia.